The Relative Cost-Effectiveness of Retaining Versus Accessing Air Force Pilots

Michael G. Mattock, Beth J. Asch, James Hosek, Michael Boito

Prepared for the United States Air Force

Approved for public release; distribution unlimited

For more information on this publication, visit www.rand.org/t/RR2415

Library of Congress Cataloging-in-Publication Data is available for this publication.
ISBN: 978-1-9774-0204-2

Published by the RAND Corporation, Santa Monica, Calif.

Cover: Getty Images/Fanatic Studio.

Preface

The research discussed in this report was conducted for a project to develop an analytic capability for determining the efficient amount of special and incentive (S&I) pays for rated officers in the U.S. Air Force (USAF) given the cost of producing additional trained officers, the cost of S&I pay to retain those officers, and those officers' retention behavior in response to changes in S&I pay. To accomplish this goal, the research estimated the cost of training pilots by flying platform and broke down the cost by training stage. It extended and estimated RAND's dynamic retention model for Air Force pilots by specialty and platform, such as fighter pilots and bomber pilots, and ran simulations that showed the effects that changes in S&I pays have on retention cost and the trade-offs between training cost and retention cost by specialty and platform. The research provided information to help the Air Force make efficient decisions about retention incentives for specific pilot specialties.

The purpose of this report is to document the methods, findings, and conclusions of the project, including the finding that, from a personnel cost perspective, it is more cost-effective for the USAF to increase S&I pays and retain pilots than to expand the training pipeline to sustain a given pilot inventory.

The research reported here was sponsored by the Military Force Management Policy Directorate; Manpower, Personnel and Services; Headquarters, U.S. Air Force, and conducted within the Manpower, Personnel, and Training Program of RAND Project AIR FORCE as part of a fiscal year project "Cost–Benefit Analysis for Rated Special and Incentive Pays."

RAND Project AIR FORCE

RAND Project AIR FORCE (PAF), a division of the RAND Corporation, is the U.S. Air Force's federally funded research and development center for studies and analyses. PAF provides the Air Force with independent analyses of policy alternatives affecting the development, employment, combat readiness, and support of current and future air, space, and cyber forces. Research is conducted in four programs: Strategy and Doctrine; Force Modernization and Employment; Manpower, Personnel, and Training; and Resource Management. The research reported here was prepared under contract FA7014-16-D-1000.

Additional information about PAF is available on our website: www.rand.org/paf/

This report documents work originally shared with the U.S. Air Force on November 15, 2017. The draft report, issued on February 16, 2018, was reviewed by formal peer reviewers and U.S. Air Force subject-matter experts.

Contents

Figures

Tables

Summary

Aviation Bonus (AvB) and Aviation Incentive Pay (AvIP) are the two special and incentive (S&I) pays that the U.S. Air Force (USAF) uses to influence pilot retention and sustain the size of the pilot force. These pays have received increased attention in recent years for two reasons. First, changes in the commercial-airline industry have led to concerns about pilot retention in the USAF and about whether AvB and AvIP are high enough. The commercial-airline industry will be hiring pilots in increasing numbers in the next ten years to replace its aging pilot workforce, and recent changes to Federal Aviation Administration regulations on pilot rest and on the number of flying hours needed to qualify for an airline transport pilot certificate could add to this demand. Second, AvIP and AvB have become discretionary programs under Department of Defense Instruction 7730.67 (Office of the Under Secretary of Defense for Personnel and Readiness, 2016), so the USAF must annually justify its budgets for and usage of AvIP and AvB.

An alternative to relying on retaining current pilots to sustain the size of the pilot force is to access and train new pilots. But expanding the pilot training pipeline is costly, so, ultimately, the USAF faces a trade-off between increasing pilot accessions and incurring greater training cost or increasing pilot retention and incurring higher S&I pay cost, as well as the higher personnel cost of a more senior force. For a given-size pilot force, the determination of the efficient level of S&I pay and whether it is more cost-effective to train new pilots or retain those already in the force depends on the cost of training; personnel cost, including the cost of S&I pay; and, importantly, how responsive pilot retention behavior is to increases in AvB and AvIP.

To ensure that AvB and AvIP use USAF resources efficiently to sustain the pilot force, the USAF asked RAND Project AIR FORCE to develop an analytic capability for determining the efficient amount of S&I pay for a rated officer career field given the cost of producing an additional trained and adequately experienced rated officer, as well as the retention behavior of rated officers. This report summarizes our analysis in response to this request.

To develop the capability, we estimated the cost of training a USAF pilot for different specialties and platforms, including fighter pilots, bomber pilots, and mobility pilots. We also extended RAND's dynamic retention model (DRM) for USAF pilots in general, summarized in Mattock, Hosek, et al., 2016, to these pilot specialties, and we estimated separate models for each specialty using longitudinal data on the careers of USAF pilots, provided by the Defense Manpower Data Center (DMDC). We then used these model estimates together with the DRM computer code to simulate the retention effects of alternative S&I pay amounts and the cost trade-off between accessing and training more pilots on the one hand and using S&I pays to retain them beyond their active duty service commitments on the other. Throughout, we assumed that civilian pilot wages did not need to change in response to changes in AvB for the labor market to clear.[1]

Findings

Training Cost per Basic Qualified Pilot Is Generally Quite High

In our study, we estimated the cost of training a basic qualified pilot in the USAF, by aircraft type, following the cost methodology recommended by the Air Education and Training Command (AETC). The method involved retrieving Air Force Total Ownership Cost data on aircraft operating and support costs and usage, determining fixed and variable costs, and calculating cost per flying hour for each stage of training. The training stages include initial flight screening, undergraduate pilot training, and training to become initially qualified in an aircraft at the formal training unit. For fighter pilots, the training stages also include an introduction to fighter fundamentals. To calculate the cost of training a basic qualified pilot, we multiplied the cost per flying hour by the number of flying hours from the pilot syllabus then summed the costs at each state. Table S.1 shows our estimates of total cost of training a basic qualified pilot.

We found that training cost per pilot is significant across platforms. The cost of training a basic qualified fighter pilot ranges from $5.6 million for an F-16 pilot to $10.9 million for an F-22 pilot. Bomber pilot training cost is also high, ranging from $7.3 million for a B-1 pilot to $9.7 million for a B-52 pilot. Costs for training transport pilots and mobility pilots are somewhat lower, but still considerable, ranging from $1.1 million for a C-17 pilot to $2.5 million for a C-130J pilot. Training cost per pilot for command, control, intelligence, surveillance, and reconnaissance operations (e.g., the RC-135) is about $5.5 million.

[1] A market clears if the price (in this case, the civilian pilot wage) is such that supply equals demand. In this case, the labor market clears if the supply of pilots equals the demand for them.

Table S.1. The Total Cost of Training a Basic Qualified Pilot, by Aircraft Platform, in Fiscal Year 2018 Dollars

Aircraft	Cost per Pilot
A-10	5,961,000
B-1	7,338,000
B-2	9,891,000
B-52	9,688,000
C-130J	2,474,000
C-17	1,097,000
C-5	1,397,000
F-15C	9,200,000
F-15E	5,580,000
F-16	5,618,000
F-22	10,897,000
F-35A basic[a]	10,167,000
F-35A transition[b]	9,467,000
KC-135	1,196,000
RC-135	5,447,000

SOURCES: Air Force Total Ownership Cost data; Air Combat Command, 2014a, 2014b, 2015, 2016, and 2017; AETC, 2012, 2013, 2014a, 2014b, 2014c, 2015a, 2015b, and 2015c; Air Force Global Strike Command, 2012 and 2016; Air Mobility Command, 2016.
[a] A student must have graduated from introduction to fighter fundamentals.
[b] For pilots previously qualified on F-35 or other fighter or attack aircraft.

Analysis Shows That Retaining Pilots Is More Efficient Than Accessing New Ones

As AvB increases, the force becomes more experienced, so basic pay, retirement cost, the cost of allowances, and the cost of the AvB increase. As an example, Figure S.1 shows the simulated steady-state fighter pilot retention profile (number of pilots by years of service [YOSs]) when the AvB cap is $25,000 or $35,000 per year of obligated service and pilots are paid at the cap. When AvB is higher (at $35,000 in our model, indicated by the green line), fewer accessions are required to sustain the pilot inventory, and more pilots stay longer, beyond 20 YOSs. Because of the increase in experience as AvB increases, per capita personnel cost increases, where we computed per capita cost based on the total pilot inventory. However, per capita training cost is lower because fewer pilots are trained and those who are trained also stay longer because of the higher AvB. If the decrease in training cost offsets the increase in personnel cost (including the AvB cost), per capita pilot cost falls as AvB increases. On the other hand, if the decrease is not offsetting, per capita pilot cost will rise as AvB increases. Thus, a priori, it is ambiguous whether it is less costly to retain or train more pilots to sustain a given force size.

Figure S.1. Simulated Steady-State Fighter Pilot Retention Profile, Aviation Bonus Cap of $25,000 or $35,000

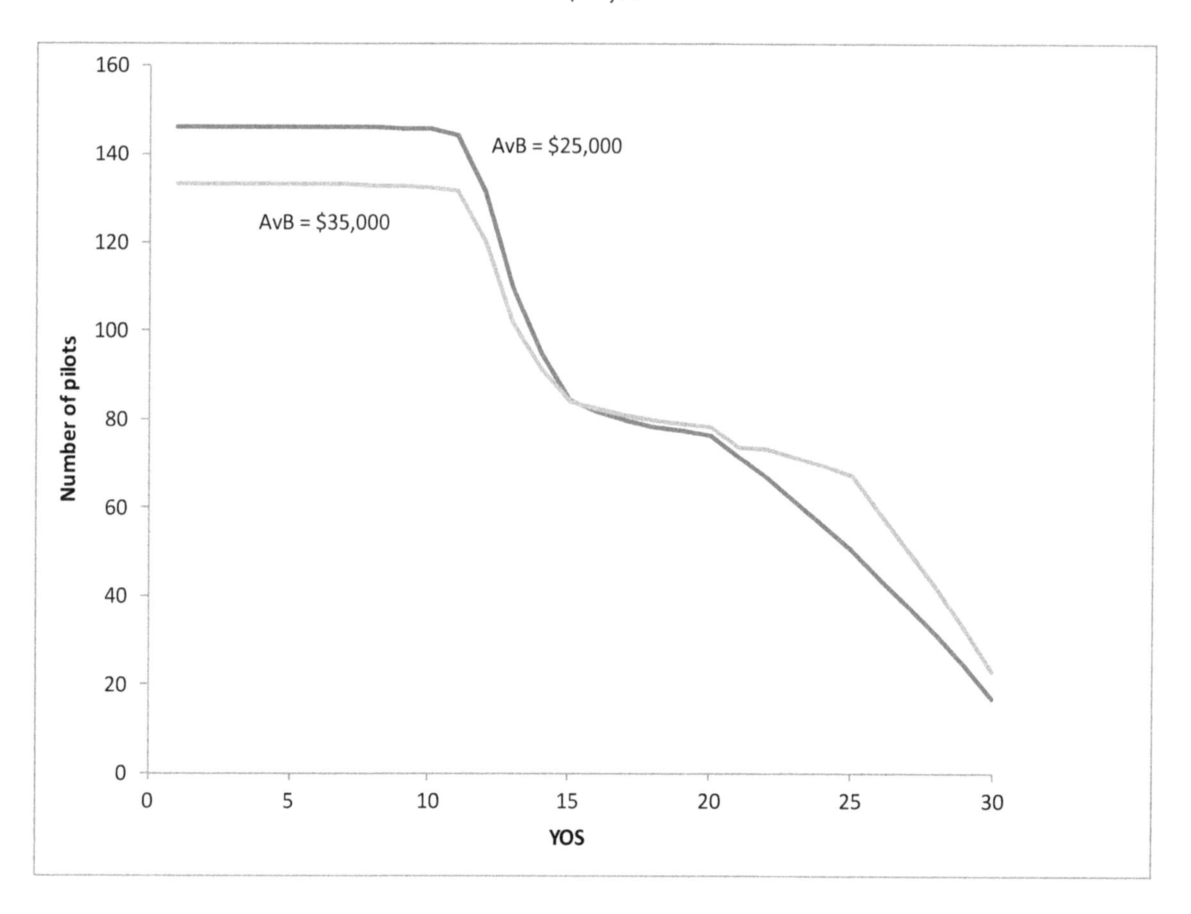

SOURCES: DRM simulation results; DMDC personnel data.

We based the simulated retention profiles shown in Figure S.1 on an estimated DRM for fighter pilots. We also estimated models for other specialties for which we had sufficient data, including bomber pilots, mobility pilots, reconnaissance pilots, and trainer pilots. The parameter estimates were generally significant, and the estimated models fit the data quite well. The major-airline hiring coefficient in the fighter pilot model was significantly different from that of every other model in which this coefficient was significantly different from 0. The other parameter estimates were somewhat similar across models, and often not statistically different between models. We used the estimates to develop a simulation capability to simulate how pilot retention varies with AvB in the steady state, resulting in, for example, the retention profiles shown in Figure S.1. We used the estimated retention effects of AvB together with the training cost and personnel cost estimates to calculate the per capita cost as AvB increased for each pilot specialty. Specifically, we considered how per capita personnel cost varied as the AvB cap increased from $0 to $95,000 per year, although the current AvB cap is $35,000 per year. Personnel cost includes basic pay, basic allowance for housing, basic allowance for subsistence, the tax advantage of receiving allowances tax free, retirement accrual calculated as a percentage of pay, and AvB and AvIP costs. This personnel cost takes into account the change in the experience mix of the force as AvB increases. It also includes the cost of training pilots using the training cost in Table S.1. We computed per capita cost as the total cost of the pilot inventory divided by the size of the pilot inventory, where the pilot inventory is specified as the median inventory between fiscal year (FY) 2008 and FY 2017. For example, the median inventory for fighter pilots from FY 2008 to FY 2017 was 3,050.

Our key finding is that, for each of the pilot specialties we considered—fighter, mobility, bomber, reconnaissance, and trainer—increasing AvB to increase retention was more efficient than expanding the training pipeline to sustain a given pilot inventory, over the range of AvB amounts we considered. Figure S.2 illustrates the result for fighter pilots. The figure shows per capita cost as the AvB cap increases when training costs per pilot are $5.6 million (the blue line) and $10.9 million (the red line). These training costs per fighter pilot were at the low and high ends of the training cost range for fighter pilots, respectively. In both cases, per capita cost declines as AvB increases. The decline is nonlinear and is steeper when AvB is between $0 and $40,000 and less steep thereafter. The decline is nonlinear because retention is more responsive to increases in AvB at lower levels of AvB. As AvB increases, pilots with higher taste for the USAF are already retained, and, at the margin, it is more difficult to retain pilots with a given increase in AvB.

Figure S.2. Cost per Fighter Pilot, by Aviation Bonus Cap, When Training Cost per Pilot Ranges from $5.6 Million to $10.9 Million, in Fiscal Year 2018 Dollars

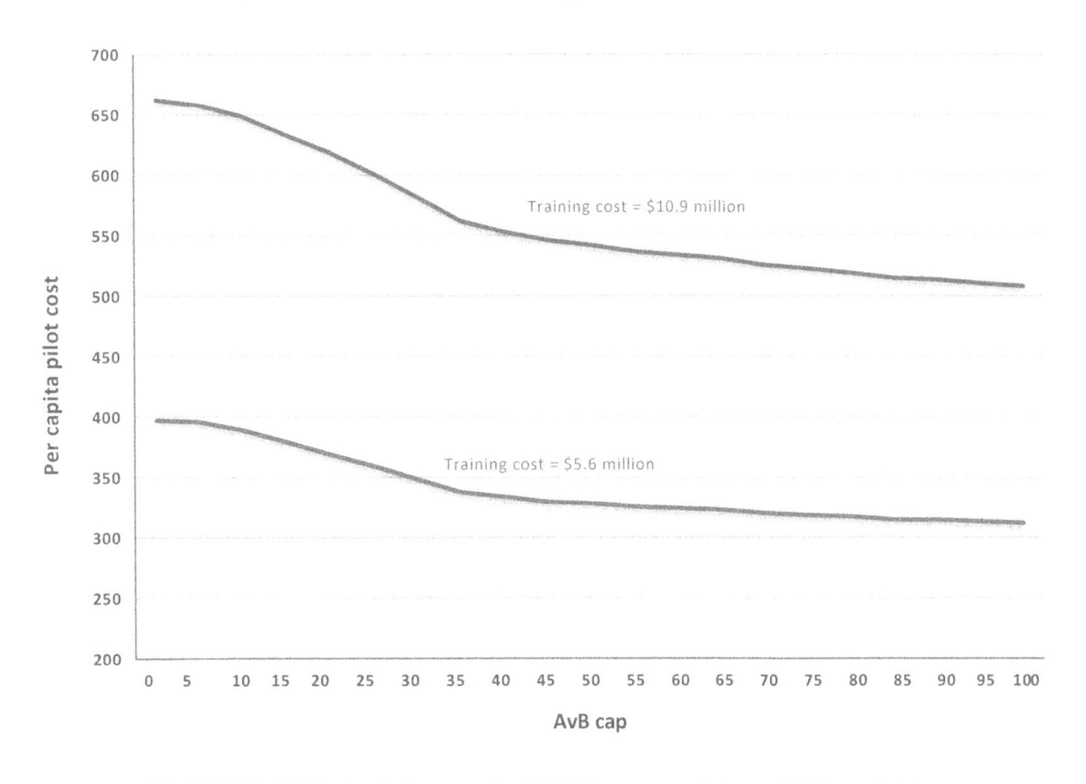

SOURCES: DRM simulation results; DMDC personnel data; AETC cost data.

The reason that increasing AvB is more efficient than increasing accessions is that pilot training cost is so high. We found that, were pilot training cost lower—$1 million, in the case of fighter pilots—per capita cost would not continue to decline as AvB increases, and, in fact, the efficient level of AvB (i.e., the level that minimizes per capita cost) would be $45,000 per year of additional commitment.

From the standpoint of minimizing steady-state per capita pilot cost, our results indicate that it is efficient to increase the AvB cap to at least $100,000 to retain midcareer fighter pilots in the steady state. That said, the experience mix of the fighter pilot force would be more senior at an AvB of $100,000 than when AvB is lower. Figure S.1 shows this general point for the case in which the AvB cap increases more modestly, from $25,000 to $35,000. Although the total size of the fighter inventory is the same in both cases—3,050 pilots—the higher AvB leads to a more experienced force, a force that might be more experienced than the USAF desires. Thus, whether it is efficient to continue to increase AvB also depends on USAF requirements for the most-experienced pilots.

The retention and cost results pertain to the steady state when the entire force is under a higher level of AvB. We further developed and used the DRM simulation capability to consider the effects on retention during the transition years, when either AvB or accession is increased as a means of increasing overall pilot strength. Because of the time required to train an experienced pilot, the number of experienced pilots in the inventory could differ substantially in the transition years if the inventory is increased because of higher AvB versus if the USAF increases number of pilots in the training pipeline (i.e., accesses more pilots).

We found evidence to support this hypothesis. The share of pilots who are experienced is larger in the early transition years, when S&I pay, rather than the size of the training pipeline, is increased, although the experience mix of pilots is more senior in the steady state. Thus, insofar as the USAF would like to increase the number of experienced pilots more quickly in the transition years, increasing S&I pay is a better tool for doing so.

Conclusions

Our analysis shows that, from a personnel cost perspective, it is more cost-effective for the USAF to increase S&I pay and retain pilots than to expand the training pipeline to sustain a given pilot inventory. In short, these results are driven by the high cost of training pilots, which is determined by the current curricula, policies, capacities, and technologies used to train USAF pilots. Changes in how the USAF trains pilots could decrease training cost, although the decrease would have to be sufficiently large to reverse our results and make expanding the training pipeline more cost-effective than expanding AvB. Future research should explore where training cost savings might be found (e.g., perhaps through alternative training technologies) without jeopardizing USAF capability and readiness.

Acknowledgments

We are grateful for the support of our project sponsor, Maj Gen Brian T. Kelly, former director, Military Force Management Policy. We also benefited from the input and support of our project monitor, Lt Col William MacDonald, former chief, Rated Force Policy, Military Force Policy Division, as well as Col William D. Fischer, chief, Military Force Policy Division, and Earl Anders, an analyst in Air Education and Training Command/Financial Management and Comptroller. We are also grateful for the help we received from several of our RAND colleagues, including John A. Ausink, William W. Taylor, Tara L. Terry, Paul Emslie, and Anthony Lawrence. Our report benefited from the reviews we received from Ellen M. Pint and Alexander D. Rothenberg of RAND. Finally, we would like to thank Raymond E. Conley, director of the Manpower, Personnel, and Training Program within RAND Project AIR FORCE.

Abbreviations

AC	active component
ADSO	active duty service obligation
AETC	Air Education and Training Command
AFSC	Air Force Specialty Code
AFTOC	Air Force Total Ownership Cost
ARP	Aviator Retention Pay
ATP	airline transport pilot
AvB	Aviation Bonus
AvIP	Aviation Incentive Pay
C2ISR	command, control, intelligence, surveillance, and reconnaissance
DoD	U.S. Department of Defense
DMDC	Defense Manpower Data Center
DRM	dynamic retention model
FAPA	Future and Active Pilot Advisors
FTU	formal training unit
FY	fiscal year
IFF	introduction to fighter fundamentals
IFS	initial flight screening
MAH	major-airline hire
N/A	not applicable
NCP	normal cost percentage
O&S	operations and support
SD	standard deviation
S&I	special and incentive
USAF	U.S. Air Force
YAS	year of aviation service
YOS	year of service

1. Introduction

The U.S. Air Force (USAF) relies on the Aviation Bonus (AvB) and Aviation Incentive Pay (AvIP) programs, two programs providing special and incentive (S&I) pays, to influence USAF pilot retention. AvB and AvIP can offset internal conditions or external market forces that could decrease retention, including changes in the demand for pilots in the airline industry. In recent years, these retention incentives have been an important tool for the USAF to respond to the increasing civilian hiring of pilots by major airlines. AvIP, also called *flight pay*, is paid to an officer on aviation duty as a monthly amount ranging from $150 to $1,000, depending on the officer's years of aviation service (YASs) (Office of the Under Secretary of Defense for Personnel and Readiness, undated).[2] AvB is a financial incentive paid to experienced pilots who have completed their active duty service obligations (ADSOs) to continue on active duty for a specific period of additional service chosen by the pilot. The USAF can pay up to $35,000 per year of obligated service for each AvB agreement through 25 YASs. For example, a pilot who has completed 11 years of service (YOSs) could receive up to $455,000 in bonuses for staying an additional 13 years.

An alternative to retaining pilots to sustain the size of the pilot force is to access and train new pilots. But expanding the training pipeline can be costly, at $11 million for a fifth-generation fighter, for example, according to Lt Gen Gina M. Grosso in 2017 testimony to the U.S. House of Representatives Committee on Armed Services Subcommittee on Military Personnel (Woody, 2017). In addition, the USAF could face constraints in absorbing pilots into units and ensuring that they have completed the requisite flying hours to become experienced pilots.

AvIP and AvB are discretionary programs under Department of Defense Instruction 7730.67 (Office of the Under Secretary of Defense for Personnel and Readiness, 2016), so it is important for the USAF to have an analytical basis to develop and justify its budgets for and usage of AvIP and AvB and to have a capability to ensure that it is allocating resources efficiently between retaining experienced rated officers and training new ones. This capability should be able to consider the cost and benefits of using AvIP and AvB to sustain the retention of pilots versus accessing and training new ones in the long run, as well as in the short run. For example, because of the time required to train new pilots, during surge periods in demand or when external pilot opportunities expand dramatically, increasing pilot retention might be more cost-effective than accessing new pilots even if, in the long run, expanding the training pipeline might be more cost-effective. Because training costs, as well as retention behavior, differ across pilot specialties and flying platforms, the capability needs to be developed for each rated specialty.

[2] Historically, these have been denoted as Aviation Continuation Pay or Aviation Career Continuation Pay and Aviation Career Incentive Pay, respectively.

The research summarized in this report was motivated by the question of what the efficient level of AvB is for each pilot specialty to meet and maintain a needed inventory level given the cost of training pilots in that specialty and platform. The objective of the project was to develop an analytic capability for determining the efficient amount of S&I pay for a rated officer career field given the cost of producing an additional trained and adequately experienced rated officer, as well as the retention behavior of rated officers. Answering this question and meeting this objective required estimates of training and other personnel costs, as well as estimates of the responsiveness of pilot retention to changes in AvB and AvIP. The study team estimated the cost of training a USAF pilot for different specialties and platforms following the costing methodology recommended by Air Education and Training Command (AETC) and using data inputs gathered from AETC and other USAF sources. We used the training cost estimates together with RAND Project AIR FORCE's dynamic retention model (DRM) to simulate the retention effects of alternative offerings of S&I pays, including AvB and AvIP, and the cost trade-offs between accessing and training more rated officers on the one hand and using S&I pays to retain rated officers beyond their active duty service commitments on the other.[3] We extended the DRM to include the requisite data inputs to allow for specialty and platform differences in rated officer retention and to simulate the effects that changes in S&I pay had on officer retention and cost. The data inputs came from Defense Manpower Data Center (DMDC) personnel data.

Chapter 2 presents a conceptual overview of the trade-offs between achieving a given pilot inventory with higher retention incentives on the one hand and more accessions on the other. This overview provides context for the simulated cost trade-offs between pilot retention and accessions shown later in the report. Chapter 3 discusses the training cost estimation methodology and pilot training cost estimates, by platform. In Chapter 4, we present an overview of the DRM and the estimation of separate models for specific pilot specialties and platforms: fighter pilots, bomber pilots, mobility pilots, reconnaissance pilots, and trainer pilots.[4] We present an overview because the DRM for pilots is presented in detail in Mattock, Hosek, et al., 2016. Chapter 5 shows the analytic results on the cost trade-offs between retaining more pilots and expanding the training pipeline, holding pilot inventory constant. We offer closing thoughts in Chapter 6.

---

[3] The DRM was developed and is discussed in Mattock, Hosek, et al., 2016, building on an earlier version, discussed in Mattock and Arkes, 2007. Other reports on recent studies of USAF pilot retention include McGee, 2015, and Sweeney, 2015.

[4] Although the USAF trains and employs pilots of remotely piloted aircraft, we did not include them in this analysis because the lack of historical data on conditions that affect retention for these pilots prevented us from modeling their behavior like we have done for other pilot communities.

2. Conceptual Trade-Off Between Retaining Pilots and Expanding the Training Pipeline

The focus of this study was on providing estimates of the costs and benefits of sustaining the inventory of pilots for different platforms in the USAF and offsetting the increase in major airlines' hiring by increasing retention incentives—specifically, AvB—versus expanding the training pipeline. Expanding AvB versus expanding pilot accessions and the number of pilots trained have different implications for cost and the experience mix of the force. This chapter provides a conceptual framework for these costs and benefits. We provide estimates in subsequent chapters.

Figure 2.1 shows the simulated steady-state retention profile for fighter pilots under two scenarios, holding fighter pilot force size constant at 3,050. In one scenario, AvB is set at \$25,000 per year; in the second, AvB is \$35,000 per year. The retention profile shows the number of pilots, by YOSs, for which we performed the simulations using the DRM estimates for fighter pilots discussed in Chapter 4. A key finding shown in Figure 2.1 is that accessions are lower when AvB is higher, and the experience mix of the force increases. Specifically, steady-state accessions are 133, versus 147 when AvB is higher in the figure, while the inventory of pilots with 15 or more YOSs increases because retention increases as a result of the higher incentive to remain in the USAF.

The differences in accessions and experience mix as AvB increases, holding force size constant, have cost implications. The per capita cost C of a pilot is given by the formula

$$C = \frac{PersonnelCost + \left(TrainingCostperPilot \times PilotsTrained\right)}{PilotForceSize},$$

where personnel cost for the pilot force includes the cost of cash compensation, such as basic pay, as well as retirement accrual cost that is computed as a percentage of basic pay. Training cost per pilot is the cost per pilot trained and accounts for washouts. To obtain total cost of pilot training, we multiplied the per-pilot training cost by the number of pilots trained. Training cost per pilot varies with pilot specialty and platform, which we discuss in Chapter 3.

Table 2.1 summarizes the cost implications of this formula. The inventory of pilots is sustained when AvB is lower and accessions are higher (the upper-left quadrant in the table) or when AvB is higher and accessions are lower (the lower-right quadrant). Otherwise, the force size increases (upper-right quadrant) or the force size decreases (lower-left quadrant).

Figure 2.1. Simulated Steady-State Fighter Pilot Retention Profile, Aviation Bonus Cap of $25,000 or $35,000

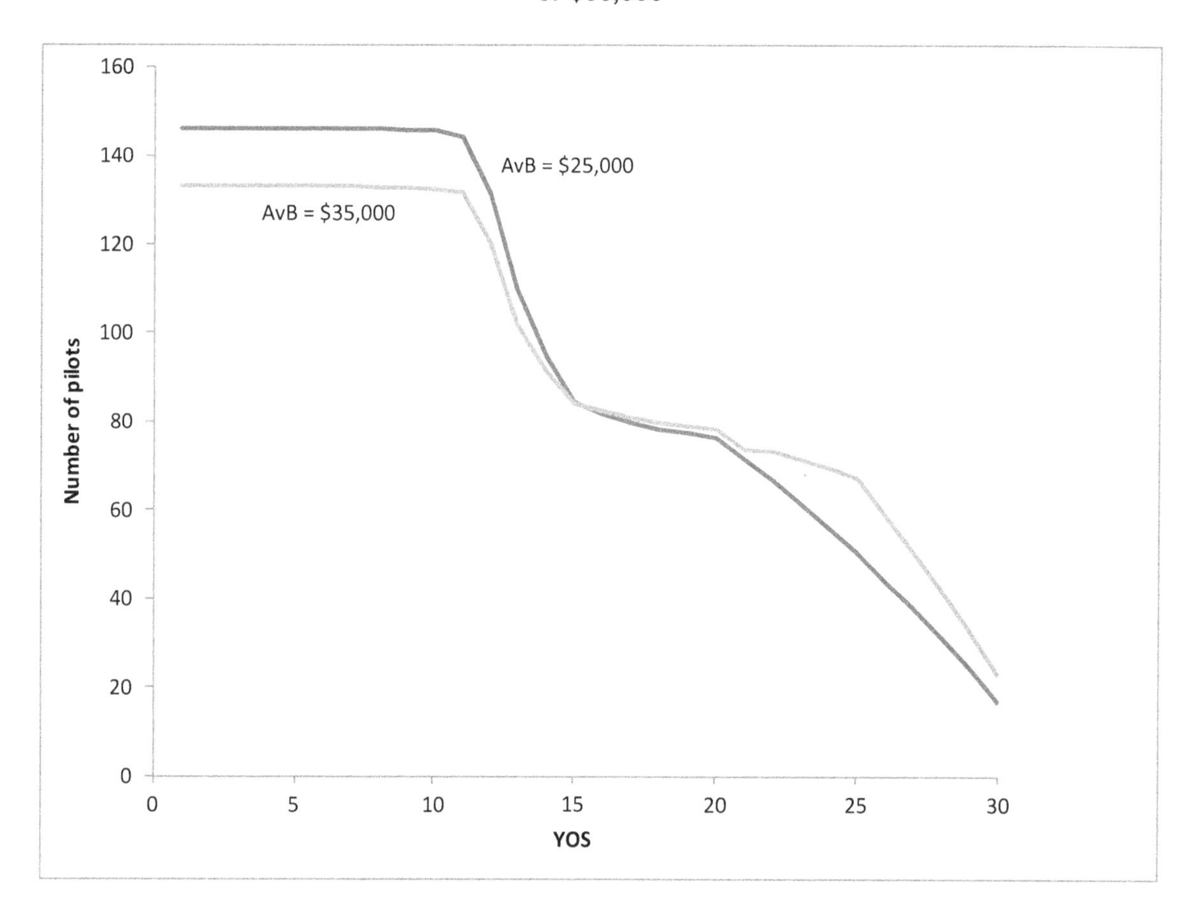

Table 2.1. Conceptual Framework: Implications of Calculations of the Cost of Training Pilots

	Lower AvB	Higher AvB
More accessions	• A less experienced force • Lower pay bill • Higher training costs (force size constant)	• N/A (larger force)
Fewer accessions	• N/A (smaller force)	• A more experienced force • Higher pay and AvB bill • Lower training costs (force size constant)

NOTE: N/A = not applicable. We see cost increases due to both higher AvB and higher regular military compensation (pay) costs due, in turn, to an increase in the seniority of the force in the lower-right quadrant, and not so much in the upper-left quadrant.

When AvB is lower and accessions are higher, the force is less experienced, as illustrated in Figure 2.1 for the case of AvB of $25,000 or $35,000. A less experienced force is less costly in terms of cost of basic pay and allowances that increase with grade, such as the basic allowances for housing and for subsistence. Retirement accrual cost is also lower for a less experienced force because the basic pay bill is lower. On the other hand, having more accessions means needing an expanded training pipeline, so the total cost of training pilots increases. Additionally, there might be capacity constraints that limit the production and absorption of pilots into units in which they can acquire necessary flying hours. Expanding capacity could mean capital expenditures on additional aircraft. The figure does not list the cost of expanding capacity to produce and absorb more pilots, but this could be an important additional cost.

On the other hand, when AvB is higher and accessions are lower, the force is more experienced, as shown in Table 2.1. A more experienced force is also more productive insofar as more-experienced personnel are more capable of performing their duties and have a broader array of capabilities. Because fewer accessions are needed to sustain a given force size, training cost is lower, and, because a given accession stays longer when AvB is higher, the USAF gets a larger return on the training investment. Still, the more-experienced force means that the compensation bill, including retirement accrual cost, is higher for a given force size.

It is clear from this framework that, a priori, there is ambiguity about which is more costly to sustain a given force size: retaining more current pilots or training more new ones. As AvB increases, personnel cost increases because the force is more experienced, but total training cost is lower because fewer pilots are trained. If the decrease in the latter offsets the increase in the former, overall total cost decreases, and, given that the pilot force size is held constant, per capita pilot cost, C, falls. But if the reverse is true, per capita cost increases as AvB increases. One possibility is that per capita cost falls over a range of AvB but then increases beyond this range. That is, at some level of AvB, per capita cost is minimized. A minimum is more likely to be observed when training cost per pilot is relatively low because, in this case, the decrease in training cost will be smaller, while personnel cost (including AvB) will increase as the cost of retaining additional pilots increases.

In Chapter 5, we show simulations of the per capita cost of pilots for different specialties and platforms for the range of per-pilot training costs we estimate in Chapter 3 and using the occupational community–specific pilot retention models we estimated for Chapter 4.

3. The Cost of Training Pilots

Is it less expensive to train or to retain? If training is inexpensive, the answer is almost certainly "to train"; if training is dear, the answer is likely to be "to retain"—up to a point. To anchor the analysis, then, we needed to have a firm grasp of training cost. Then we could balance the cost of training pilots with the cost of retaining them.

This chapter presents estimates of the cost of training USAF pilots to the status of basic qualified pilot for piloted aircraft. The chapter begins with an overview of the methodology and provides a brief description of the stages of training. It is useful to understand the training stages because different aircraft are used for different stages, giving rise to different operations and support (O&S) costs associated with each aircraft that we included in the cost estimates. The chapter then describes the cost considered in the computation of training cost and included in the calculations, the methodology, and the logic behind the methodology used to estimate training cost. As part of this discussion, we identify the different cost elements that enter into the calculation of the marginal cost and describe our rationale for treating them as fixed or variable over a multiyear period. We then discuss the sources of data used to estimate training cost. The chapter then moves to a table displaying estimates of the marginal cost of training subject to current training syllabi, training technology, and training capacity. These estimates are used in our policy simulations that change AvB subject to a given pilot force size. In the simulations, marginal cost is assumed to be constant at its point estimate. Finally, the chapter concludes with a discussion of why training Air Force officers as basic qualified pilots is costly. We used these estimates in our policy simulations that change AvB subject to a given pilot force size. In the simulations, we assumed marginal cost to be constant at its point estimate.

Overview of the Methodology

The methodology we used for estimating USAF pilot training cost generally follows that used by financial management personnel at AETC (Anders, 2017a). We first computed the total cost of operating and supporting each USAF aircraft we considered, averaged over three years (2014 through 2016), and a three-year average of hours flown by each of these aircraft, thereby allowing us to compute cost per flying hour for each aircraft. We then estimated the number of flying hours required for each stage of training, using pilot course syllabi provided by the USAF. Given the requisite flying hours for each stage and a cost per flying hour, we could then estimate the training cost at each stage and the total across stages. The sum provided an estimate of the training cost for a basic qualified pilot for each aircraft. We describe the data inputs for these calculations in the remainder of this chapter.

Stages of Pilot Training

All stages of training use classrooms, training devices or simulators, and flying instruction. We focus our discussion of cost on the flying training because it is the costliest part of training, but the data we employed included all the training cost.

The first stage of pilot training is initial flight screening (IFS). The USAF conducts IFS in Pueblo, Colorado, to assess each candidate's aptitude for flying. Most, or roughly 1,200 per year, of these candidates will become pilots or combat system officers of manned aircraft. Combat system officers have Air Force Specialty Codes (AFSCs) 12xx and are rated, nonpilot officers. They serve as the mission commanders for electronic warfare, navigation, and weapon systems on a variety of USAF aircraft, including bomber, fighter, mobility, and reconnaissance aircraft. An additional roughly 400 candidates per year are screened for training on piloting remotely piloted aircraft.

The second stage of pilot training depends on the platform. Graduates of flight screening who will pilot remotely piloted aircraft receive simulator and classroom training, then proceed to formal training units for their aircraft. For pilots who will fly piloted aircraft, graduates of IFS proceed to primary undergraduate pilot training, where they learn basic flying skills. The aircraft used for the flying portion of primary flying training is the T-6, a single-engine two-seat turboprop.

The third stage, after primary flying training, is advanced undergraduate pilot training in one of three tracks. Students train in a different aircraft in each track, as shown in Table 3.1.

Table 3.1. Specialized Undergraduate Pilot Training Tracks

Specialized Undergraduate Pilot Training Track	Aircraft Flown in Advanced Training
Airlift and tanker	T-1 twin-engine jet; three-person crew
Bomber and fighter	T-38 twin-engine supersonic jet; two-person crew

SOURCES: Logistics, Installations and Mission Support—Enterprise View, undated; Air Combat Command, 2014a, 2014b, 2015, 2016, and 2017; AETC, 2012, 2013, 2014a, 2014b, 2014c, 2015a, 2015b, and 2015c; Air Force Global Strike Command, 2012 and 2016; Air Mobility Command, 2016.

After graduating from advanced undergraduate pilot training, all but the fighter track graduates proceed to formal training units (FTUs) for initial qualification in the aircraft to which they will be assigned. Fighter pilots proceed from advanced undergraduate pilot training to an introduction to fighter fundamentals (IFF) course in the T-38, and then to the FTUs for their aircraft.

Table 3.2 summarizes the stages of training after IFS and the aircraft flown at each stage for the kinds of pilots for which we present cost in this report. A basic qualified pilot will have proceeded through IFS and each stage of training in the table.

Table 3.2. Aircraft Flown at Each Stage of Training for Selected Types of Pilots

	Undergraduate			
Type of Pilot	**Primary**	**Advanced**	**IFF**	**FTU**
Fighter	T-6	T-38	T-38	A-10, F-15, F-16, F-22, or F-35
Bomber	T-6	T-38		B-1, B-2, or B-52
Mobility, tanker, reconnaissance, or special operations	T-6	T-1		C-17, C-130J, KC-135, or RC-135

SOURCES: Logistics, Installations and Mission Support—Enterprise View, undated; Air Combat Command, 2014a, 2014b, 2015, 2016, and 2017; AETC, 2012, 2013, 2014a, 2014b, 2014c, 2015a, 2015b, and 2015c; Air Force Global Strike Command, 2012 and 2016; Air Mobility Command, 2016.

Pilot Training Cost

Our objective was to estimate the cost of training more USAF pilots to the status of basic qualified pilot. We considered two broad categories of costs related to pilot training: indirect cost, which encompasses the cost of the O&S of the bases at which training is conducted, and direct cost, or cost clearly tied to the O&S of the aircraft used in training. Direct cost includes the cost of personnel to maintain and otherwise support the aircraft, as well as the cost of instructor and student pilots and other instructor personnel.

Indirect Cost Associated with Pilot Training

Pilot training through the stage of basic qualified pilot is conducted at many locations. AETC conducts undergraduate pilot training at Columbus, Laughlin, Randolph, Sheppard, and Vance Air Force Bases. Basic flight qualification training at FTUs is conducted at an additional large number of bases.

The O&S cost for these bases includes such functions as medical operations, personnel support, base services, utilities, maintenance of property, administration, housing, morale, welfare, and recreation and support all units, personnel, and infrastructure at the base. This cost is considered indirectly related to pilot training. Because the USAF was able to produce 1,236 pilots as recently as fiscal year (FY) 2012, we assumed that the base infrastructure would be sufficient for planned pilot production. And because indirect base cost tends to be relatively insensitive to changes in the number of personnel supported (Laverson, 2000), we assumed that there would be little increase in indirect base O&S cost and excluded that cost from our calculation of marginal cost.

Direct Cost of Pilot Training

The direct cost of pilot training is primarily for flying training but also includes classroom and training device or simulator training.

Direct O&S cost of U.S. Department of Defense (DoD) systems is reported in a standard cost element structure. The cost elements in this structure are largely self-explanatory and indicate the kinds of costs incurred in training and considered in this analysis. The structure has six major elements, with further levels of indentation. The cost element structure, for the elements for which we have cost data, is shown in the "Cost Element: Second Level" column in Table 3.3.

Table 3.3. Operations and Support Cost Element Structure

Cost Element: Top Level	Cost Element: Second Level	Description
1.0 Unit-level personnel	1.1 Operations personnel 1.2 Unit-level maintenance personnel 1.3 Other unit-level personnel	Includes crew, maintenance, and staff personnel
2.0 Unit operations	2.1 Operating material 2.2 Support services 2.3 Temporary duty 2.4 Transportation	Primarily for fuel but also includes the cost of munitions used in training and other expenses at the unit level
3.0 Maintenance	3.1 Consumable materials and repair parts 3.2 Depot-level reparables 3.4 Depot maintenance 3.6 Interim contractor support 3.7 Contractor logistics support	Includes consumable and reparable parts replaced at the unit, as well as aircraft and engine overhauls and similar depot maintenance, whether provided by government or contractor organizations
4.0 Sustaining support	4.3 Sustaining and systems engineering 4.4 Program management 4.6 Data and technical publications 4.7 Simulator operations and repair 4.8 Other sustaining support	Includes simulator operations, which are an important part of the training syllabus for pilots
5.0 Continuing system improvements	5.1 Hardware modifications 5.2 Software maintenance	Includes the cost of equipment purchased with procurement funds and installed on aircraft when they are operational
6.0 Indirect support	6.1 Installation support 6.2 Personnel support 6.3 General training and education	Includes the share of the installation indirect support cost discussed in the preceding section that can be attributed to the weapon system. For aircraft used for training, the indirect support element of general training and education includes the cost of instructors and students and is a much larger proportion of total O&S cost than for nontraining aircraft.

SOURCE: Data are from Air Force Total Ownership Cost (AFTOC), homepage, last modified July 26, 2018.

By far, the costliest part of pilot training is flying, which is a direct cost. The cost of flying training has two parts. One part is hours flown by the student pilot. The second part is hours flown by instructor pilots (when they are in separate aircraft) and hours flown by other direct support aircraft. Other direct support includes flying by adversaries and other support aircraft, such as tankers. Requirements for sorties by direct support aircraft are identified in the pilot training syllabus. The syllabus is explained in "Sources of Data" later in this chapter.

Costs Included in Our Estimates

Having described the nature of the costs associated with pilot training, we next explain the rationale for including or excluding those costs in our calculations. The objective of our analysis affected the choice of costs to include. We were interested in the marginal cost of training the additional number of pilots needed at various levels of S&I pay. The FY 2018 budget request identified a need to train an additional 136 pilots of piloted aircraft per year. We anticipated that the USAF would need to train this higher level of pilots for several years. Our analysis of retention incentives was intended to examine the cost and benefit of increased retention of pilots in this yearly range of numbers over many years.

Determining Fixed Versus Variable Direct Cost

Two issues had a significant effect on our choice of costs to include in our calculation of direct training cost. The first issue was the fixed or variable nature of the direct cost of flying activity and whether to include both of these kinds of costs. The second issue was whether to include the cost of direct support aircraft.

In Table 3.4, we indicate our assessment of whether each element is fixed or variable with respect to increased flying activity and pilot production *over a multiyear period*, along with our rationale for each assessment.

For aircraft used only for training, we believe that all direct O&S cost should be included in the cost of training students, including those costs that do not vary closely with flying activity. The fixed cost elements have no other purpose than training, so it is reasonable to apportion them to this purpose. Thus, for the T-1, T-6, and T-38 aircraft used in undergraduate pilot training, we counted all direct O&S cost as part of the cost of training.

Aircraft flown in FTUs could potentially be used to perform missions other than training and therefore present a somewhat different case. One could argue that the maintenance personnel, for example, gain experience on a platform that could, at some point in the maintainer's career, be applied to a similar aircraft with a mission other than training. Or one could argue that the modifications to an aircraft in an FTU might one day be useful in a combat mission if the aircraft becomes combat coded. Under these assumptions, some or all of the cost of these elements for training aircraft might be apportioned to a purpose other than training.

We argue that the FTU is a permanent step in the stages of training to produce a basic qualified pilot. The most accurate representation of the cost of this stage was to include all its cost, rather than only costs that vary with usage. However, for a sensitivity analysis, we also calculated only the variable cost of aircraft at FTUs and present those results at the end of this chapter.

Table 3.4. Assessment of Fixed or Variable Nature of Training Cost

O&S Cost Element	Our Assessment of Fixed or Variable with Flying Activity	Rationale
1.0 Unit-level personnel	Semivariable	Unit-level personnel are assigned to a unit for a period of time. Personnel cost is incurred for that duration and is relatively insensitive to minor changes in flying activity. However, over a multiyear period, prolonged changes in flying activity or pilot production would result in less-than-proportionate changes in the numbers of unit-level personnel. The FY 2018 budget request for additional maintenance personnel to accommodate increased undergraduate pilot production indicates that there is a relationship between the number of unit-level personnel and the number of pilots completing undergraduate pilot training.
2.0 Unit operations	Variable	Fuel and training munitions are consumed when the aircraft is used. Fuel cost is the bulk of unit operations cost. This cost would change nearly proportionally to changes in flying and pilot production.
3.0 Maintenance	Variable	The cost of consumable and reparable parts is closely related to flying activity. The cost of depot maintenance correlates less closely with flying, largely because it reflects the deferred costs of accumulated usage but is driven partially by usage over long periods. We expect that maintenance cost in total would vary closely with changes in pilot production over a multiyear period.
4.0 Sustaining support	Mostly fixed	Sustaining support cost is largely insensitive to changes in flying activity because most tasks must be accomplished regardless of changes in usage or fleet size.
5.0 Continuing system improvements	Fixed	Continuing system improvements or modifications are made to provide capability for a fleet and not in response to changes in flying activity. System improvement cost would change proportionally with changes in fleet size.
6.0 Indirect support	Mostly variable	The element of indirect support is further broken down into installation support, personnel support, and general training and education. Installation and personnel support cost tends to be a small portion of total O&S cost. As explained in the previous section, indirect support cost in general is not sensitive to changes in direct personnel. However, for the training aircraft fleets, the cost of students and most instructors is reflected in the indirect element of general training and education. For these fleets, in contrast to aircraft not used for initial training, the indirect cost for general training and education is a large proportion of total O&S cost and varies directly with pilot production.

SOURCE: AFTOC data.

Rationale for Including Direct Support Cost

As noted above, some of a training aircraft's flying hours and direct O&S cost are due to sorties flown by student pilots. But a significant additional amount includes flying by instructor pilots. Direct support sorties and cost also include adversary aircraft in combat training sorties and other support aircraft, such as tankers. These additional sorties are specified in the training syllabus for each aircraft.

However, sorties that are not specified in the training syllabus are those required for instructor pilots to achieve and maintain proficiency in the trainers in which they will instruct. For example, for the T-1 in FY 2018, training for instructor pilots is estimated to account for

31 percent of total T-1 fleet flying hours. In the next section, we explain how we ensured that we captured this cost for instructor proficiency flying in undergraduate pilot training aircraft.

Methodology for Calculating Training Cost per Student for Each Aircraft Used in Training

We used different methodologies to calculate training cost per student, depending on the information we had available. The methodologies differed for the T-1 and T-6 aircraft used in undergraduate pilot training, for the T-38 aircraft used in advanced undergraduate pilot training, and for all other aircraft at the FTUs. The essential feature of the methodologies was to capture the aircraft direct O&S cost per graduate for each aircraft flown in training. In "Sources of Data" later in this chapter, we describe the data sources used.

For the T-1 and T-6 aircraft used in undergraduate pilot training, we obtained yearly direct O&S cost for them from AFTOC. The cost includes the cost of hours flown by students and instructor pilots. We obtained the costs for FYs 2014 through 2016 and normalized the costs to constant FY 2018 dollars.

We obtained the number of T-1 graduates per year from AETC and divided T-1 costs by the number of graduates for FYs 2014 through 2016 to obtain an average cost per graduate. This method ensured that we captured the cost of flying hours for instructor pilot training.

We estimated the number of T-6 graduates per year by adding the numbers of T-1 and T-38 graduates per year (numbers provided by AETC Operations via our project sponsor [McDonald, 2017]) and an estimated 200 combat system officers per year who fly the T-6 in their training. We divided T-6 costs for FYs 2014 through 2016 by the estimated number of T-6 graduates to obtain an average cost per graduate.

We used a different method to estimate T-38 cost per graduate. The T-38 is used for advanced undergraduate pilot training; IFF; and other purposes, such as an adversary aircraft role against other fighters during training. The T-38 can be used for different purposes at the same base. We were not able to isolate the T-38 cost for each purpose and divide the cost by the number of graduates like we did for the other training aircraft. Instead, for the T-38 fleet, we obtained costs and flying hours for FYs 2014 through 2016 but excluded the cost and flying hours of aircraft operated by Air Combat Command, reasoning that these aircraft were not used in an undergraduate pilot training role. We calculated the average cost per flying hour for these T-38 aircraft. We multiplied the cost per flying hour by the number of flying hours specified in the training syllabi for advanced undergraduate pilot training and IFF, plus flying hours for instructor pilot proficiency training, which are not specified in the training syllabi. We did not have information on the requirement for instructor pilot proficiency training for the T-38 and assumed 25 percent of the student flying hours, which likely errs low.

For all aircraft at formal training units, we read the training syllabus for each aircraft to determine the total sortie and flying-hour requirement per student, including the hours for direct support aircraft. For the relevant aircraft except F-35, we obtained the annual costs and flying hours for FYs 2014 through 2016, normalized the costs to FY 2018 dollars, and calculated an average cost per flying hour. We multiplied the number of flying hours per student for each type of aircraft and respective costs per flying hour to determine the cost per student at these stages of training.

For the F-35, we used cost and flying hours for FY 2016 only. F-35 cost per flying hour has been decreasing as more aircraft are fielded and more flying hours are flown each year. It is likely that the F-35 cost per flying hour will continue to decrease from the FY 2016 level. Given trends to date and estimates of the F-35's O&S cost over its life cycle, we estimated that its cost per flying hour could decrease from the FY 2016 cost by as much as 10 to 15 percent. However, given the uncertainty inherent in estimates, we used the FY 2016 cost per flying hour in our calculation.

Assumptions and Clarifications Regarding Flying Hours Required in Syllabi

Information on flying hours is needed for computing FTU cost of training; we obtained that information from the training syllabi provided by the USAF. In most cases, the number of flying hours required was straightforward; in other cases, we had to make assumptions about what to include or had to adjust the number of flying hours provided.

In our calculations, we included the refly rate (the average percentage of training sorties that must be repeated) specified in the syllabi.

Several of the syllabi specify sorties for direct support aircraft, primarily for adversary aircraft sorties (for combat aircraft training) and for tanker support. The syllabi do not specify the mission, design, and series of these direct support aircraft, although they usually indicate whether similar or dissimilar aircraft are desired in an adversary role. In these instances, we made assumptions that erred on the side of lower estimated cost. We assumed that dissimilar adversary aircraft were T-38s, which had a lower cost per flying hour than any of the fighter aircraft. We assumed that tanker sorties were provided by KC-135s and further assumed that one tanker sortie supported four training aircraft sorties.

The C-17 and KC-135 syllabi specify that each instructor have more than one student. Therefore, we adjusted the flying-hour costs so that the costs were per graduate.

There was a syllabus for F-35A initial qualification pilots and another syllabus for F-35A transition and requalification pilots who had previously qualified in the F-35A or who had previously flown other fighter or attack aircraft. For the transition and requalification pilots, we used the hours in training track 1, and, in calculating the cost to produce this pilot, assumed that an F-16 pilot had transitioned and included the cost of training a basic qualified F-16 pilot.

Sources of Data

We obtained the cost of the contract for IFS from a DoD press release on the award of the contract (DoD, 2017). We used approximate numbers of students screened per year from the performance work statement for the recompetition for the contract.

All other cost data came from the AFTOC system. AFTOC is the USAF's official repository of O&S cost. We used cost in the standard cost element structure for O&S cost shown in Tables 3.3 and 3.4 and supplemented this information with queries of the AFTOC database in other formats. AFTOC also has aircraft programmatic data by aircraft mission, design, and series, including flying hours and whether the hours are for operational or training missions, aircraft inventory, and personnel assigned by rank and grade. AFTOC does not have programmatic data on simulator or classroom usage or throughput.

We obtained additional programmatic, usage, and maintenance data from the Logistics, Installations and Mission Support—Enterprise View system on the Air Force Portal. This source provided the purpose codes of aircraft by location and their usage.

Finally, by reading training syllabi, we obtained information on the flying instruction required for students at each stage of training. In the list of references at the end of this report, we list the syllabi. The syllabi specify sorties flown by students, instructors, and direct support aircraft.

Results

Table 3.5 displays the cost at each stage of training. For FTUs, it shows variable and total costs of the stage. The variable cost reflects a narrow definition of that cost and includes only maintenance and unit operations costs. Total FTU cost includes variable costs plus all other direct aircraft O&S cost, which consists of unit-level personnel, sustaining support, modification, and the portion of indirect cost logically tied to the weapon system. We reiterate that we believe that total FTU cost is most appropriate for calculating the marginal cost of producing an additional trained pilot but display the variable cost as a sensitivity analysis. The two rightmost columns show the sum of the costs of the stages to produce a basic qualified pilot. The second-to-last column sums the stages to produce a basic qualified pilot and includes the variable FTU cost, and the rightmost column sums the stages to produce a basic qualified pilot and includes total FTU cost.

Table 3.5. Cost at Each Stage of Pilot Training and Variable and Total Cost of Training a Basic Qualified Pilot, in Thousands of Fiscal Year 2018 Dollars

Aircraft	IFS	T-6 or Undergraduate Remotely Piloted Aircraft Training	T-1	T-38	T-38 IFF	FTU Variable	FTU Total	Basic Pilot, Variable	Basic Pilot, Total
A-10	19	370		1,232	379	1,949	3,960	3,949	5,961
B-1	19	370		1,232		3,171	5,717	4,792	7,338
B-2	19	370		1,232		4,384	8,270	6,005	9,891
B-52	19	370		1,232		4,427	8,067	6,047	9,688
C-130J	19	370	469			761	1,616	1,618	2,474
C-17	19	370	469			158	239	1,016	1,097
C-5	19	370	469			230	539	1,088	1,397
F-15C	19	370		1,232	329	4,255	7,250	6,205	9,200
F-15E	19	370		1,232	387	2,213	3,571	4,221	5,580
F-16	19	370		1,232	387	1,688	3,610	3,696	5,618
F-22	19	370		1,232	329	4,751	8,947	6,700	10,897
F-35A basic	19	370		1,232	329	4,102	8,218	6,052	10,167
F-35A transition	19	370		1,232	329	1,936	3,848	7,392	9,467
KC-135	19	370	469			219	339	1,076	1,196
RC-135	19	370	469			2,947	4,589	3,805	5,447

SOURCES: AFTOC; flying hours from various training syllabi.

We found that training cost per pilot (shown in the rightmost column of Table 3.5) ranged from about $5.6 million for the F-16 and F-15E to about $10.9 million per pilot for the F-22. The cost of training a bomber pilot is also quite high, ranging from $7.3 million per pilot for the B-1 to $9.9 million for the B-2. Mobility pilots are less costly to train. We estimated the range of costs per pilot of $1.1 million for the C-17 to a cost of $2.5 million for the C-130J. Training cost for command, control, intelligence, surveillance, and reconnaissance (C2ISR) pilots is in the midrange, at $5.4 million for the RC-135.

In short, we found that training USAF officers to become basic qualified pilots is quite costly, especially for fighter and bomber pilots. Table 3.6 gives some insight about why cost is so high. It shows the percentage of O&S cost for FYs 2014 through 2016 for USAF aircraft overall, and for the T-38 trainer specifically, attributable to major cost elements listed in Table 3.3. The large majority of aircraft O&S cost for aircraft overall is for the first three cost elements—namely, cost at the unit level (specifically unit-level personnel, fuel, training munitions, and other operating costs at the unit level) and maintenance. The proportions of these elements change somewhat over time as fuel and other costs fluctuate (not shown), but the proportions shown for FYs 2014 to 2016 and that we included in our study and show in Table 3.6 are close to the 20-year average proportions. That said, the magnitude of the cost has changed over time, even though the proportion of costs has been relatively stable. Aircraft O&S

cost has risen faster than the rate of inflation in the past 20 years because of growth in the costs of military compensation, oil, and military equipment maintenance. Table 3.6 also shows the proportions for the T-38 trainer, the aircraft used in undergraduate pilot training. The proportions differ from all aircraft because the trainer aircraft has a higher proportion of cost for indirect support, which includes cost for instructors, and a lower proportion for unit-level personnel than average USAF aircraft.

Table 3.6. Percentage of the Total Cost of Training a Basic Qualified Pilot That Is Attributable to Major Cost Elements

Major Cost Element	All Aircraft, FYs 2014–2016	T-38 Trainer, FY 2016
1.0 Unit-level manpower	30	18
2.0 Unit operations	21	29
3.0 Maintenance	35	33
4.0 Sustaining support	1	1
5.0 Modifications	8	4
6.0 Indirect support	5	15

SOURCES: AFTOC; flying hours from various training syllabi.

One caveat is noteworthy regarding the cost estimates in Table 3.5 and the use of flying hours as an input to the cost estimates. The methodology of using flying hours to calculate FTU cost understates the training cost of aircraft that require very few flying hours (i.e., C-17, C-5, and KC-135). Compared with training for other aircraft, training for these aircraft utilizes more classroom and simulation training than flying training (that is, the ratio of simulator time to actual flying time is higher for these platforms than for any other USAF platform). Using flying hours as the basis of estimating FTU cost for these aircraft tends to understate the cost of compensation for student and instructor personnel for the duration of the entire training course.

To determine the degree of underestimation of these training programs, we divided the yearly pilot and instructor costs reported in AFTOC for each aircraft by the number of flying hours to calculate the cost per pilot and instructor per flying hour embedded in our methodology. We multiplied the cost per pilot and instructor per flying hour by the number of flying hours specified in each training syllabus to determine the pilot and instructor compensation cost of FTU training as estimated by the flying-hour methodology. We compared this cost with the cost of the pilot's and instructor's time using the course duration for each training syllabus expressed in days. We assumed a student pilot compensation at the average of the O-1 and O-2 grades, and instructors at the O-3 grade, with a ratio of two students per instructor.

The greatest extent of underestimation of student and instructor costs using the cost-per-flying-hour methodology was for the KC-135. The cost-per-flying-hour methodology produced a student and instructor cost roughly $50,000 less than an estimate based on course duration in days. The $50,000 is 15 percent of the estimate of the KC-135 FTU stage and 4 percent of the estimate for the total cost of training a basic qualified pilot.

Because the higher estimation of compensation cost did not affect the results of the cost–benefit analysis described in Chapter 4, we did not change the estimates produced by the cost-per-flying-hour methodology.

4. Dynamic Retention Model Overview, Estimates, and Model Fits

The DRM is a model of the service member's decision, made each year, to stay in or leave the active component (AC) and, for those who leave, to choose whether to participate in a reserve component and, if participating, whether to continue as a reservist. In the model, these decisions are structured as a dynamic program in which the service member seeks to choose the best career path but the path is subject to uncertainty. The model is formulated in terms of parameters that are estimated with longitudinal data on retention in the AC and participation in the reserve component, then used to see how well the estimated model fits observed retention. We use the estimated parameters in policy simulations.

We have described the DRM in earlier reports, in which we have estimated a DRM for officers and for enlisted personnel in each service (Asch et al., 2008) and for selected communities, such as military mental health care providers (Hosek et al., 2017) and USAF pilots (Mattock, Hosek, et al., 2016). In this chapter, we show coefficient estimates and model fit for selected USAF commissioned officer AFSCs corresponding to different communities of USAF pilots.

In the DRM, a set of parameters underlies the individual officer's retention decisions, and a goal of our analysis was to use individual-level data on officers' retention over their careers to estimate the parameters. An underlying assumption of the model is that people are free to choose to stay or leave active service and anticipate being able to revisit the decision to stay or leave as their careers progress. Thus, we use the DMDC's Work Experience File to track individual officers' careers from 1990 through 2000 cohorts until 2012, predating the most active period of USAF "force shaping," which resulted in some officers having to end their active careers earlier than they had anticipated.

Model Overview

In the behavioral model underlying the DRM, in each period, the service member can choose to continue on active duty, leave the military to hold a job as a civilian, or leave the military to join a reserve component and hold a job as a civilian. The member bases his or her decision on which alternative has the maximum value. The model assumes that a service member begins his or her military career in an AC.

We assumed that people differ in their preferences for serving in the military. We assumed that each officer has given, unobserved, preferences for commissioned active and reserve service and that the preferences do not change. The officer presumably has knowledge of military pay and retirement benefits, as well as civilian compensation. In each period, random shocks are associated with each of the alternatives, and the shocks affect the value of the alternative. The model explicitly accounts for individual preferences and military and civilian compensation, and, in this context, shocks represent current-period conditions that affect the value of being on active duty, being in the selected reserve and being a civilian worker (or *reserve*, for short), or being a civilian worker and not in the selected reserve (*civilian*, for short). Examples of what can contribute to a shock are a good assignment; a dangerous mission; an excellent leader; inadequate training or equipment for the tasks at hand; a strong or weak civilian job market; an opportunity for on-the-job training or promotion; the choice of location; a change in marital, dependency, or health status; the prospect of deployment or deployment itself; or a change in school tuition rates. These factors can affect the relative payoff of being in an AC, being in a reserve component, or being a civilian. We assume that the service member knows the distributions that generate the shocks, as well as the shock realizations in the current period but not in future periods.

Depending on the alternative chosen, the service member receives the pay associated with serving in an AC, working as a civilian, or serving in a reserve component and working as a civilian. In addition, the member receives the intrinsic monetary equivalent of the preference for serving in an AC or serving in a reserve component. These values are assumed to be relative to that of working as a civilian, which is set at 0.

In considering each alternative, the service member takes into account his or her current state and type. *State* is defined by whether the officer is active, reserve, or civilian and by the officer's numbers of active commissioned YOSs, reserve commissioned YOSs, age in years, and random shocks. For officers without prior enlisted YOSs, state is set to 0.

Type refers to the level of the officer's preferences for active and reserve service. The service member recognizes that today's choice affects military and civilian compensation in future periods. The service member does not know what the realizations of the random shocks will be in future periods. The expected value of the shock in each state is 0. Depending on the values of the shocks in a future period, any of the alternatives—active, reserve, or civilian—might be the best for that officer at that time. Once a future period has been reached and the shocks are realized, the service member can reoptimize (i.e., choose the alternative with the maximum value for the service member at that time). The possibility of reoptimizing is a key feature of dynamic programming models that distinguishes them from other dynamic models. In the current period, with future realizations unknown, the best the service member can do is estimate the expected value of the best choice in the next period (i.e., the expected value of the maximum). Logically, this will also be true in the next period, and the one after it, and so forth, so the model is forward-looking and rationally handles future uncertainty. Moreover, the model presumes that the service member can reoptimize in each future period, depending on the state and shocks in that period. Thus, today's decision takes into account the possibility of future career changes and assumes that the service member can reoptimize when making future decisions.

A service member can also choose to take a multiperiod contract, if offered one. A service member who chooses a longer contract can receive the associated retention bonus for more years, but the member is also "locked into" the contract such that he or she forgoes the opportunity to take advantage of better opportunities that might present themselves during the contract period. The choice is further complicated by fact that the service member cannot know the specific future conditions (e.g., assignments, flying time, deployments) that accompany different contract length choices. Thus, the contract length choice is a decision made under uncertainty.

Finally, this model uses an expected wage line that is a combination of veteran civilian nonpilot and veteran civilian pilot earnings, where earnings are weighted according to an estimated probability that an officer can find a job match to his or her taste with a major airline. (We sometimes refer to this as the *probability of being hired* by a major airline.) The probability is modeled as a function of the number of major-airline hires (MAHs) as reported by the Future and Active Pilot Advisors (FAPA) for a given year; the more MAHs, the more likely it is that an officer will be offered a position that is a good match to his or her geographic or other preferences. Specifically, we estimated the probability p that an officer can find a job match to his or her taste among major airlines as a logit function of MAH x with intercept a and slope b, where

$$p = \frac{e^{a+bx}}{1+e^{a+bx}}.$$

Special and Incentive Pays for Rated Personnel

To successfully estimate the coefficients of the DRM for a pilot AFSC, in addition to historical data on retention decisions of individual officers, regular military compensation, and the civilian opportunity wage that USAF pilots face, we needed historical data on the incentive pays available and how they might have changed over time. In estimating the model, we made the simplifying assumption that civilian pilot wages did not need to change in response to changes in USAF S&I pays for the labor market to clear.[5]

There are two main kinds of incentive pays for rated personnel: AvIP and AvB. Historically, all rated personnel have received AvIP, which has two purposes: to compensate for a career that is more hazardous than most military careers and to provide a retention incentive. AvIP provides up to $1,000 per month for a midcareer officer. (See Office of the Under Secretary of Defense for Personnel and Readiness, 2016.)

AvB is received by any rated officer who commits to a multiyear obligation, and the amount typically varies with the occupation and length of the obligation incurred. Three common options that have been offered pilots have been three-year, five-year, and until–20 YASs contract at amounts of up to the congressionally set AvB cap, which was $25,000 per year for most of the past two decades. The until–20 YASs contract offer means that the pilot will get the AvB stream

---

[5] The number of pilots separating from the regular USAF from 1990 to 2013 ranged from 600 to 1,800 per year and averaged just over 1,400 per year, or just about 11 percent of the total USAF pilot inventory of approximately 13,000. To fly for a major airline, a pilot must hold an airline transport pilot (ATP) certificate. If every separating pilot obtained an ATP certificate in his or her year of separation, they would total 13 to 28 percent of originally issued ATP certificates over those years. Because USAF pilots account for only a fraction of the potential labor supply, a partial-equilibrium approach does not seem entirely unreasonable.

of payments each year until YAS 20 if he or she stays until he or she reaches YAS 20. In addition, AvB recipients have sometimes been offered the option to take up to 50 percent of the stream of future payments up front as a lump sum, with the remainder paid out over the length of the contract. Pilots can continue in the USAF without signing up for AvB, but they receive only AvIP (that is, one receives AvIP only if not taking one of the multiyear contract options; if either option is taken, both AvIP and AvB are received). If eligible, the pilot can accept an AvB contract at a later date.

The AvB program has varied considerably over the years as USAF personnel managers have adjusted eligibility rules and the menu of contracts offered in response to outside market forces and USAF personnel retention needs. From 2000 to 2004, the USAF offered three-year, five-year, until–20 YAS, and until–25 YAS contracts with pay up to $25,000 per year for both aviators who were initially eligible and those who were beyond their initial eligibility. Initially eligible officers are those who have completed their initial active duty service commitments and are given a one-chance choice of an AvB or contract length. Beyond–initially eligible officers are those who have completed their first AvB or contract length obligations and have the opportunity to choose another AvB or contract length.

From 2005 to 2008, the USAF offered only a five-year contract at $25,000 per year, and that only to pilots who were at initial eligibility. From 2009 to 2012, the USAF expanded the portfolio of contracts offered to include, at first, retirement-eligible service members, and then those who were not retirement eligible and not currently under contractual obligation, offering three-, four-, and five-year contracts of $15,000 per year. In 2013, the USAF reintroduced the until–20 YAS option at $25,000 per year for some rated occupations. The National Defense Authorization Act for FY 2017 raised the cap on AvB from $25,000 to $35,000 per year, in addition to raising the monthly AvIP cap from $840 to $1,000.

Discussion of Parameter Estimates and Model Fit

The raw and transformed parameter estimates are shown in Tables 4.1 and 4.2, respectively, for the AFSCs for which we had sufficient data to estimate models. In Table 4.2, the transformed parameter estimates are in thousands of dollars, except for the correlation-of-taste parameter and the personal discount rate. The AFSCs we consider are fighter (11F), mobility (11A, 11M, and 11T), trainer (11K), reconnaissance or C2ISR (11R), and bomber (11B). The parameter estimates are generally of the expected sign and magnitude and are broadly consistent across AFSCs and with observed behavior. For example, communities with higher retention typically have higher estimated mean active taste at entry (e.g., bomber pilots at $19,600, shown in the rightmost column of Table 4.2), while lower-retention communities either have lower taste (e.g., fighter pilots at –$21,960) or larger uncertainty associated with the nested choice among multiyear contracts (e.g., C2ISR pilots at an astonishing $95,620). Nearly all coefficients are significantly different from 0 in the largest communities (fighter and mobility pilots); in the smaller communities (trainer, C2ISR, and bomber pilots), progressively fewer coefficients are significant. The MAH slope coefficient in the fighter pilot model was significantly different from that of every other model in which this coefficient was significantly different from 0. The other parameter estimates were somewhat similar across models, and often not statistically significantly different between models, indicating that, in future work, some pilot community samples could potentially be merged together for purposes of estimation.

The coefficients for the estimated response to major-airline hiring (FAPA MAH intercept and FAPA MAH slope) tell an interesting story, illustrated by Figure 4.1. Recall that we modeled the civilian opportunity wage for military pilots as being a weighted combination of veteran civilian nonpilot and veteran civilian pilot earnings, where earnings are weighted according to an estimated probability that an officer can find a job match to his or her taste with a major airline. Our model estimated this probability as a function of MAHs for each community; as a result, we can show how the different communities respond differently to changes in the demand for commercial-airline pilots.

The C2ISR community is by far the most responsive to changes in major-airline hiring. Pilots in this community tend to accumulate a greater number of flying hours, and at least one of the platforms currently in use requires pilots hold an ATP certificate, which is also what is required to pilot many of the platforms used by major airlines. The bomber community is by far the least responsive; this might be because bomber pilots tend to accumulate fewer flying hours than other USAF pilots, making them less competitive on the commercial market. Surprisingly, fighter pilots also seem to be among the least responsive; this could be because of lower flying hours (although at least one commercial airline, Delta, has given weight to the number of sorties flown by a pilot when it makes hiring decisions), because career development opportunities in the USAF are perceived as being better than civilian opportunities for these pilots, or because of little taste for flying a passenger aircraft after flying a fighter. The responsiveness levels of the mobility and trainer communities are similar to one another and are a close match to the overall pilot average responsiveness.

Table 4.1. Parameter Estimates and Standard Errors

Parameter	Fighter (11F)		Mobility (11A, 11M, and 11T)		Trainer (11K)		C2ISR (11R)		Bomber (11B)	
	Estimate	Standard Error	Estimate	Standard Error	Estimate	Standard Error	Estimate	Standard Error	Estimate	Standard Error
log(scale parameter, nest = κ)	6.06	0.13	6.08	0.05	5.85	0.20	5.90	0.14	6.24	0.13
log(scale parameter, alternatives within nest = λ_1)	5.35	0.14	5.27	0.08	4.99	0.33	4.90	0.27	4.83	0.27
log(scale parameter, alternatives within nest = λ_2)	2.10	0.73	2.29	0.44	3.75	0.65	4.56	0.49	1.55	0.00
log(absolute[mean active taste = μ_1])	3.09	0.73	1.18	1.73	1.15	7.48	–2.70	29.70	2.98	0.77
log(absolute[mean reserve taste = μ_2])	2.36	1.12	3.10	0.37	–2.62	17.91	2.17	2.18	1.71	2.48
log(SD active taste = σ_{11})	0.98	2.80	2.37	0.53	3.02	0.92	1.58	3.93	2.41	2.07
log(SD reserve taste = σ_{22})	5.15	0.20	5.08	0.12	4.44	0.46	4.44	0.46	3.08	0.68
atanh(taste correlation = ρ)	0.16	0.07	0.24	0.06	1.94	0.99	–0.86	0.28	0.98	1.13
log(–1 × switch cost: switch from active to civilian without fulfilling ADSO)	5.89	0.12	5.76	0.07	5.57	0.19	5.00	0.28	6.18	0.17
log(–1 × switch cost: switch from active or civilian to reserve without fulfilling ADSO)	6.84	0.14	6.47	0.07	6.21	0.23	6.24	0.30	6.49	0.15
log(–1 × switch cost: switch from civilian to reserve having fulfilled ADSO)	6.85	0.14	6.70	0.08	6.48	0.32	6.42	0.28	6.47	0.28
log(–1 × switch cost: switch from active to reserve having fulfilled ADSO)	6.47	0.15	6.39	0.08	6.13	0.33	6.19	0.28	5.90	0.33
FAPA MAH intercept = a	–2.78	0.76	–5.07	1.04	–6.07	2.46	–3.55	2.76	–5.94	4.69
FAPA MAH slope = b	0.63	0.22	1.66	0.35	1.93	0.83	1.99	0.96	1.05	1.14
Personal discount factor β (assumed)	0.94	N/A	0.94	N/A	0.94	N/A	0.94	N/A	0.94	N/A
–1 × log likelihood	2,735		5,257		923		582		365	

Parameter	Fighter (11F)		Mobility (11A, 11M, and 11T)		Trainer (11K)		C2ISR (11R)		Bomber (11B)	
	Estimate	Standard Error	Estimate	Standard Error	Estimate	Standard Error	Estimate	Standard Error	Estimate	Standard Error
N	1,208		1,717		355		211		200	

SOURCE: Parameter estimates from DMDC data on personnel entering active duty between 1990 and 2000.

NOTE: Gray shading indicates a coefficient that is significantly different from 0. SD = standard deviation. The scale parameter κ governs the shocks to the value functions for staying and for the reserve-versus-civilian nest and equals $\sqrt{\lambda^2 + \tau^2}$. The means and SDs of tastes for active and reserve service relative to civilian opportunities are estimated, as is the cost associated with leaving active duty before completing ADSO and switching from civilian status to participating in the reserves. The probability that an officer can find a job match is

$$p = \frac{e^{a+bx}}{1+e^{a+bx}},$$

where x is MAH. For these models, we assumed that the personal discount factor was 0.94.

Table 4.2. Transformed Parameter Estimates

Parameter	Fighter (11F)	Mobility (11A, 11M, and 11T)	Trainer (11K)	C2ISR (11R)	Bomber (11B)
Scale parameter, nest = κ	429.45	438.20	348.27	363.28	511.74
Scale parameter, alternatives within nest = λ_1	209.67	194.28	146.40	134.19	124.61
Scale parameter, alternatives within nest = λ_2	8.14	9.87	42.61	95.62	4.72
Mean active taste = μ_1	–21.96	3.25	–3.17	–0.07	19.60
Mean reserve taste = μ_2	–10.60	–22.14	–0.07	–8.77	5.54
SD active taste = σ_{11}	2.66	10.70	20.49	4.86	11.13
SD reserve taste = σ_{22}	172.75	160.94	84.96	84.43	21.80
Taste correlation = ρ	0.16	0.24	0.96	–0.70	0.75
Switch cost: switch from active to civilian without fulfilling ADSO	–362.54	–315.99	–262.82	–148.80	–481.65
Switch cost: switch from active or civilian to reserve without fulfilling ADSO	–935.14	–644.82	–497.82	–513.21	–657.05
Switch cost: switch from civilian to reserve having fulfilled ADSO	–945.82	–809.29	–650.14	–612.40	–648.36
Switch cost: switch from active to reserve having fulfilled ADSO	–643.23	–598.06	–457.49	–485.53	–364.44
FAPA MAH intercept = a	–2.78	–5.07	–6.07	–3.55	–5.94
FAPA MAH slope = b	0.63	1.66	1.93	1.99	1.05
Personal discount factor β (assumed)	0.94	0.94	0.94	0.94	0.94
N	1,208	1,717	355	211	200

SOURCE: DMDC personnel data.
NOTE: Transformed parameters are denominated in thousands of dollars, with the exception of the taste correlation and personal discount factor. Gray indicates a coefficient that is significantly different from 0. The scale parameter κ governs the shocks to the value functions for staying and for the reserve-versus-civilian nest and equals $\sqrt{\lambda^2 + \tau^2}$. The means and SDs of tastes for active and reserve service relative to civilian opportunities are estimated, as is the cost associated with leaving active duty before completing ADSO and switching from civilian status to participating in the reserves. The probability that an officer can find a job match is

$$p = \frac{e^{a+bx}}{1+e^{a+bx}},$$

where x is MAH. For these models, we assumed that the personal discount factor was 0.94.

Figure 4.1. Estimated Weight on Veteran Civilian Pilot Earnings as a Function of Major-Airline Hires

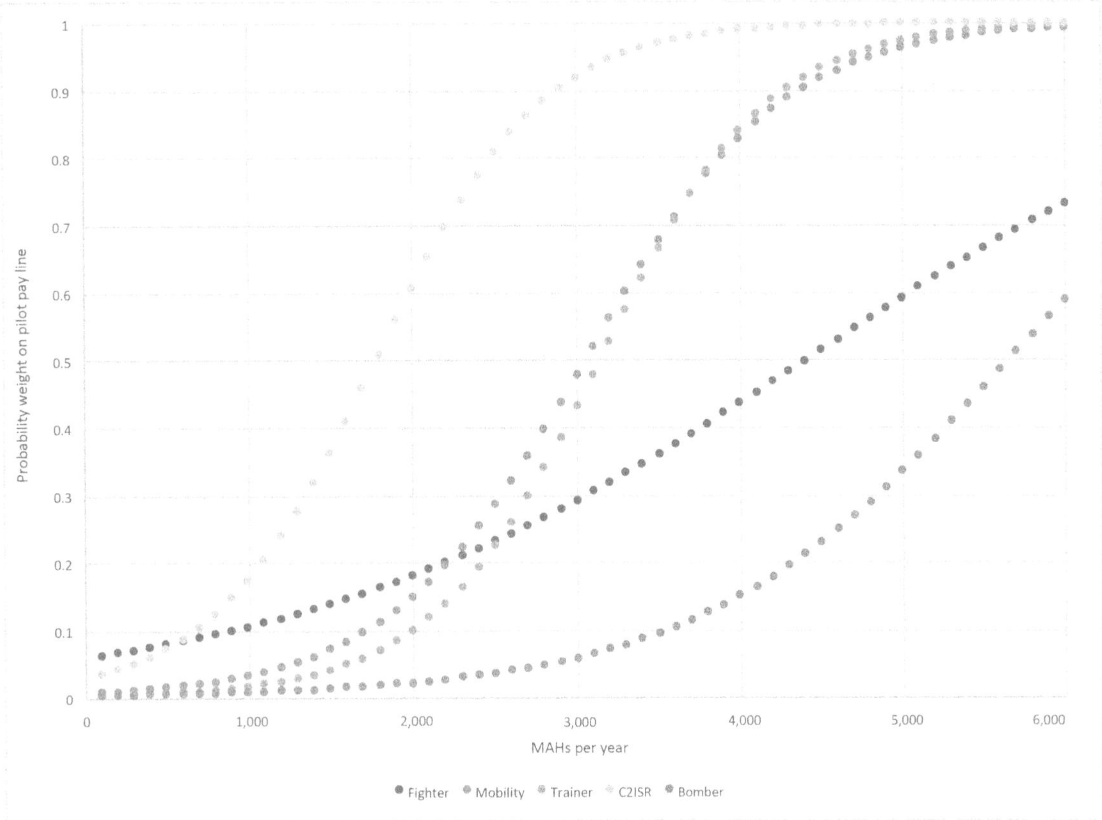

SOURCE: DMDC personnel data.

Model Fit

The model successfully replicated the characteristics of a typical pilot career for our sample cohorts, with undergraduate pilot training incurring an ADSO of eight (now ten) years, with some pilots staying only a few additional years while the remainder stayed until they vested in the retirement system at 20 years, with most leaving during the following ten years. The fit of the model was excellent across all the AFSC communities we considered, as can be seen in Figures 4.2 through 4.6. The figures show Kaplan–Meier cumulative probabilities of retention to each YOS in the AC. The black lines are the retention probabilities in the observed data, and the red lines are the predicted probabilities from the estimated models. The dotted lines are the error bands. For the most part, the red line is quite close to the black line in each figure and within the error band.

Figure 4.2. Fighter Pilot (11F) Model Fit: Observed and Simulated 11F Cohorts, 1990 Through 2000, Active Component

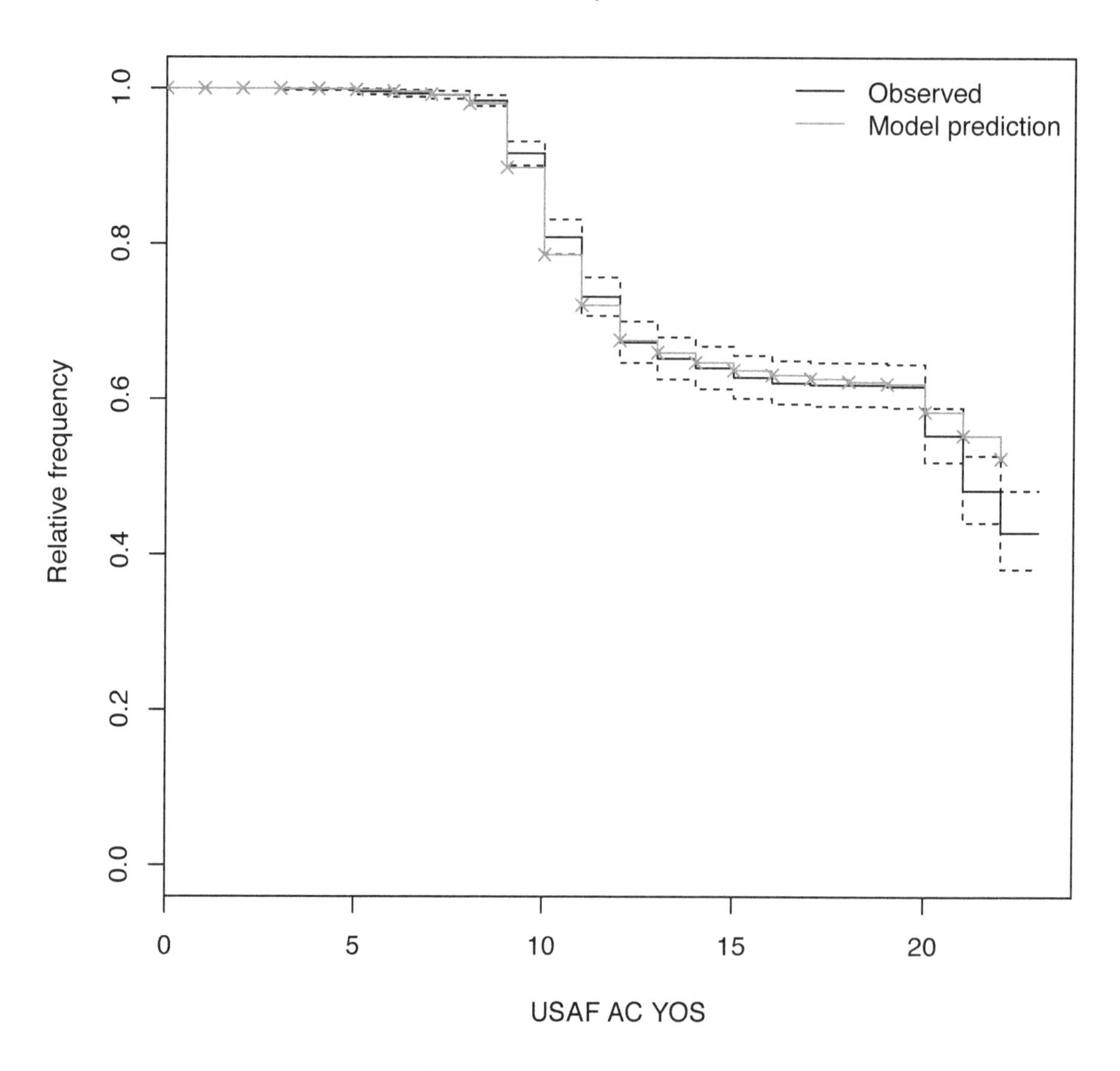

SOURCE: DMDC personnel data.
NOTE: The figure shows Kaplan–Meier cumulative probabilities of retention to each YOS in the AC. The black lines are the retention probabilities in the observed data, and the red lines are the predicted probabilities from the estimated models. The dotted lines are the error bands. For the most part, the red line is quite close to the black line and within the error band.

Figure 4.3. Mobility Pilot (11A, 11M, and 11T) Model Fit: Observed and Simulated 11A, 11M, and 11T Cohorts, 1990 Through 2000, Active Component

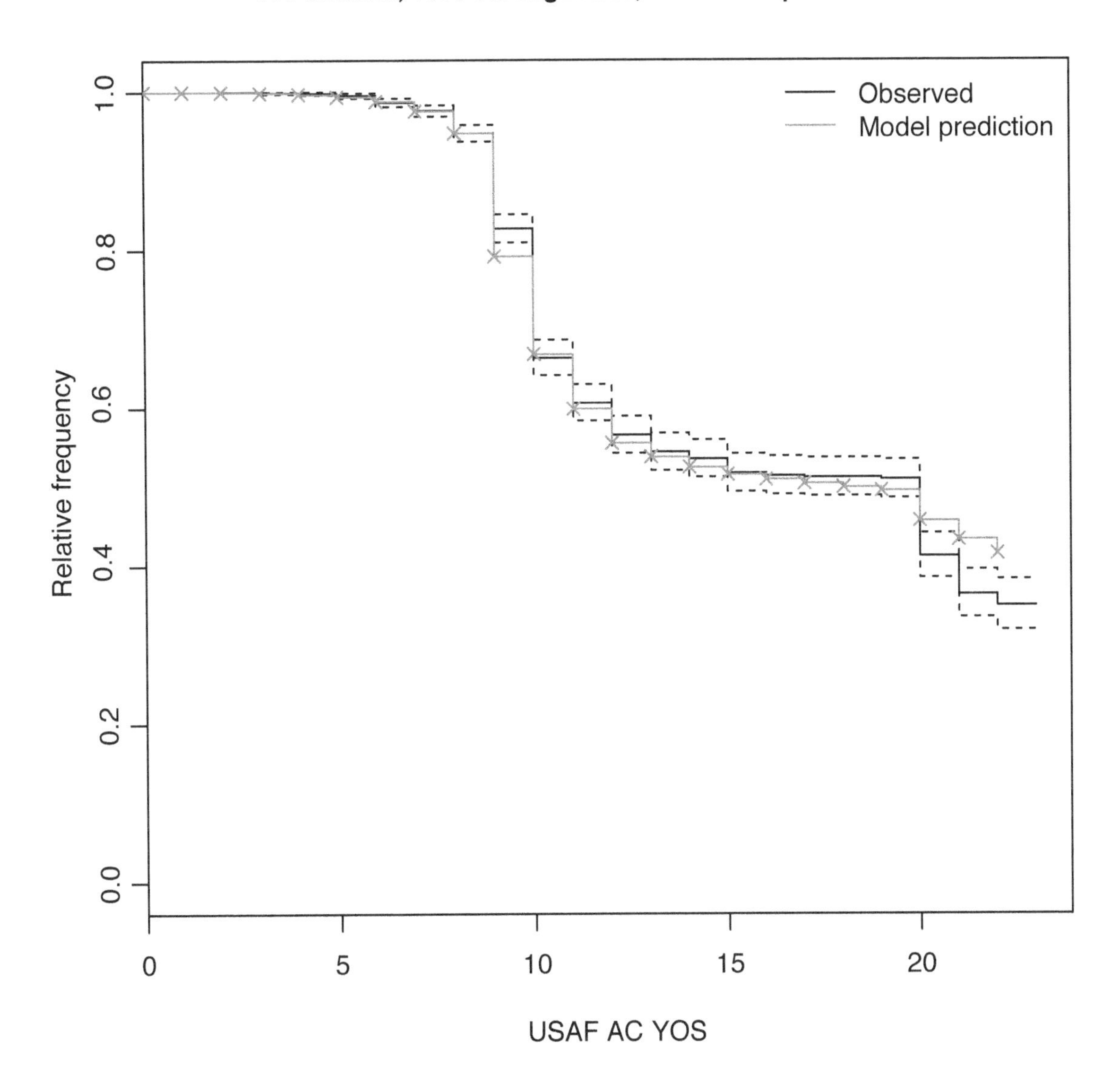

SOURCE: DMDC personnel data.
NOTE: The figure shows Kaplan–Meier cumulative probabilities of retention to each YOS in the AC. The black lines are the retention probabilities in the observed data, and the red lines are the predicted probabilities from the estimated models. The dotted lines are the error bands. For the most part, the red line is quite close to the black line and within the error band.

Figure 4.4. Trainer Pilot (11K) Model Fit: Observed and Simulated 11K Cohorts, 1990 Through 2000, Active Component

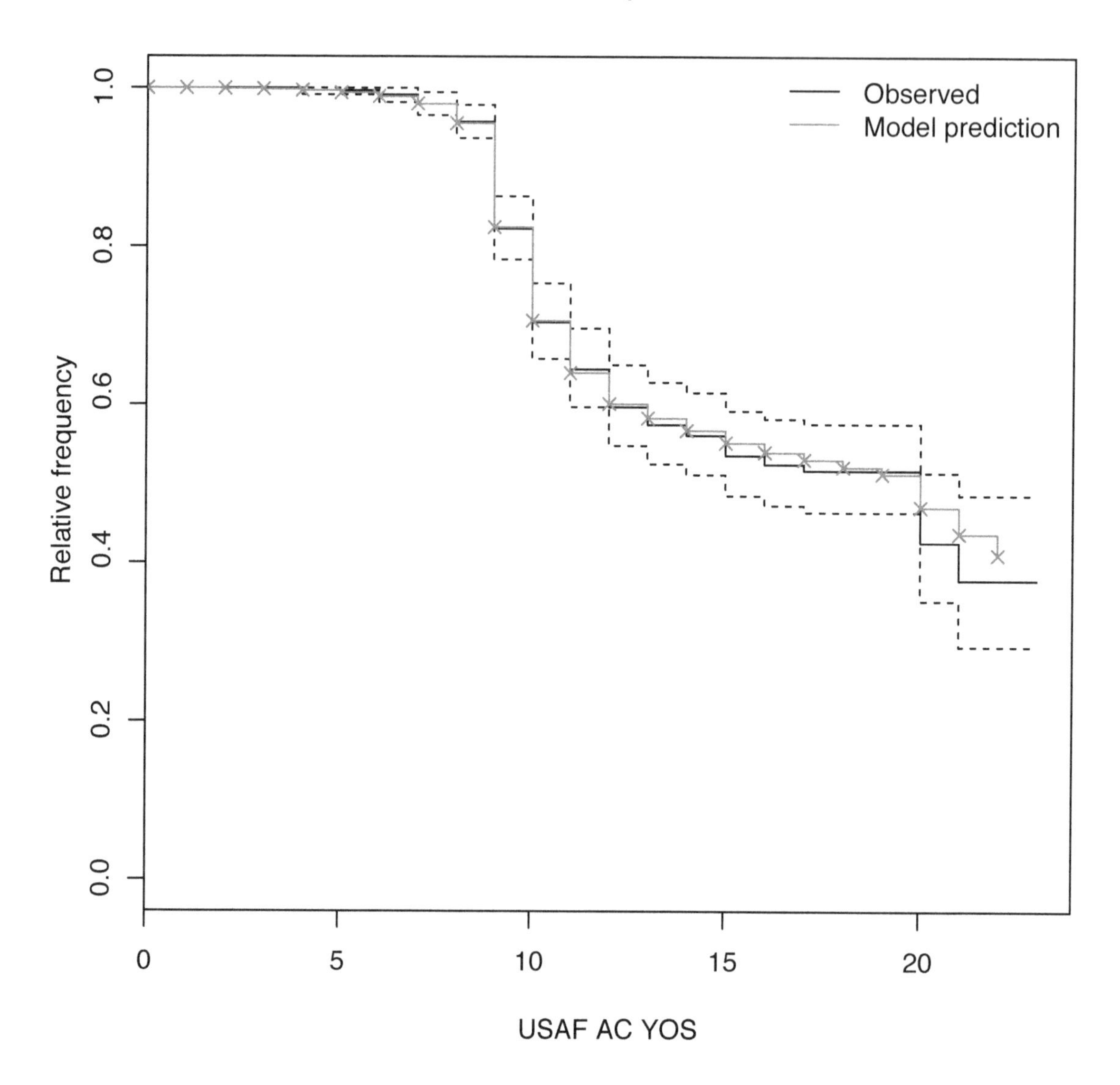

SOURCE: DMDC personnel data.
NOTE: The figure shows Kaplan–Meier cumulative probabilities of retention to each YOS in the AC. The black lines are the retention probabilities in the observed data, and the red lines are the predicted probabilities from the estimated models. The dotted lines are the error bands. For the most part, the red line is quite close to the black line and within the error band.

Figure 4.5. Command, Control, Intelligence, Surveillance, and Reconnaissance Pilot (11R) Model Fit: Observed and Simulated 11R Cohorts, 1990 Through 2000, Active Component

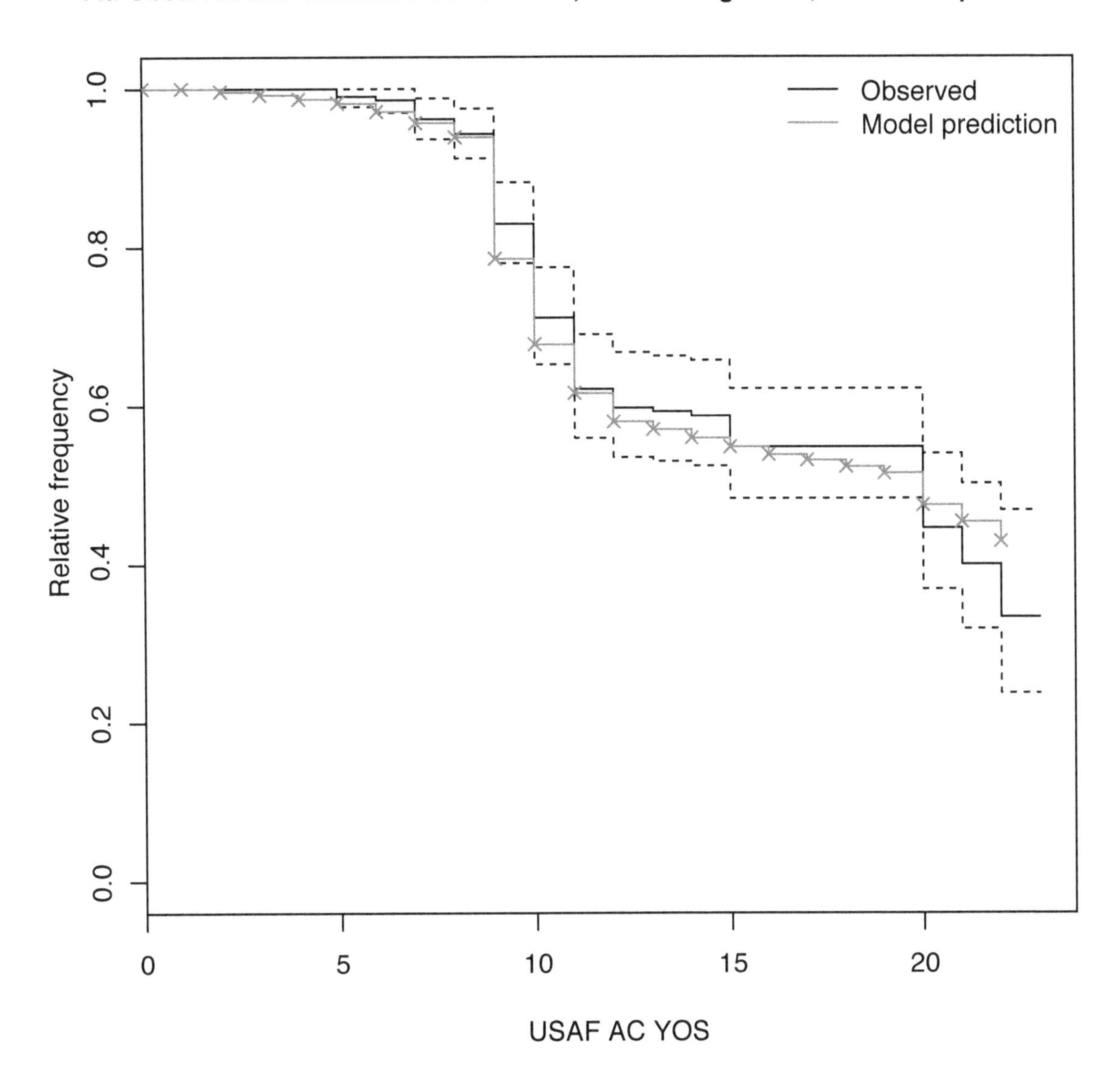

SOURCE: DMDC personnel data.
NOTE: The figure shows Kaplan–Meier cumulative probabilities of retention to each YOS in the AC. The black lines are the retention probabilities in the observed data, and the red lines are the predicted probabilities from the estimated models. The dotted lines are the error bands. For the most part, the red line is quite close to the black line and within the error band.

Figure 4.6. Bomber Pilot (11B) Model Fit: Observed and Simulated Cohorts, 1990 Through 2000, Active Component

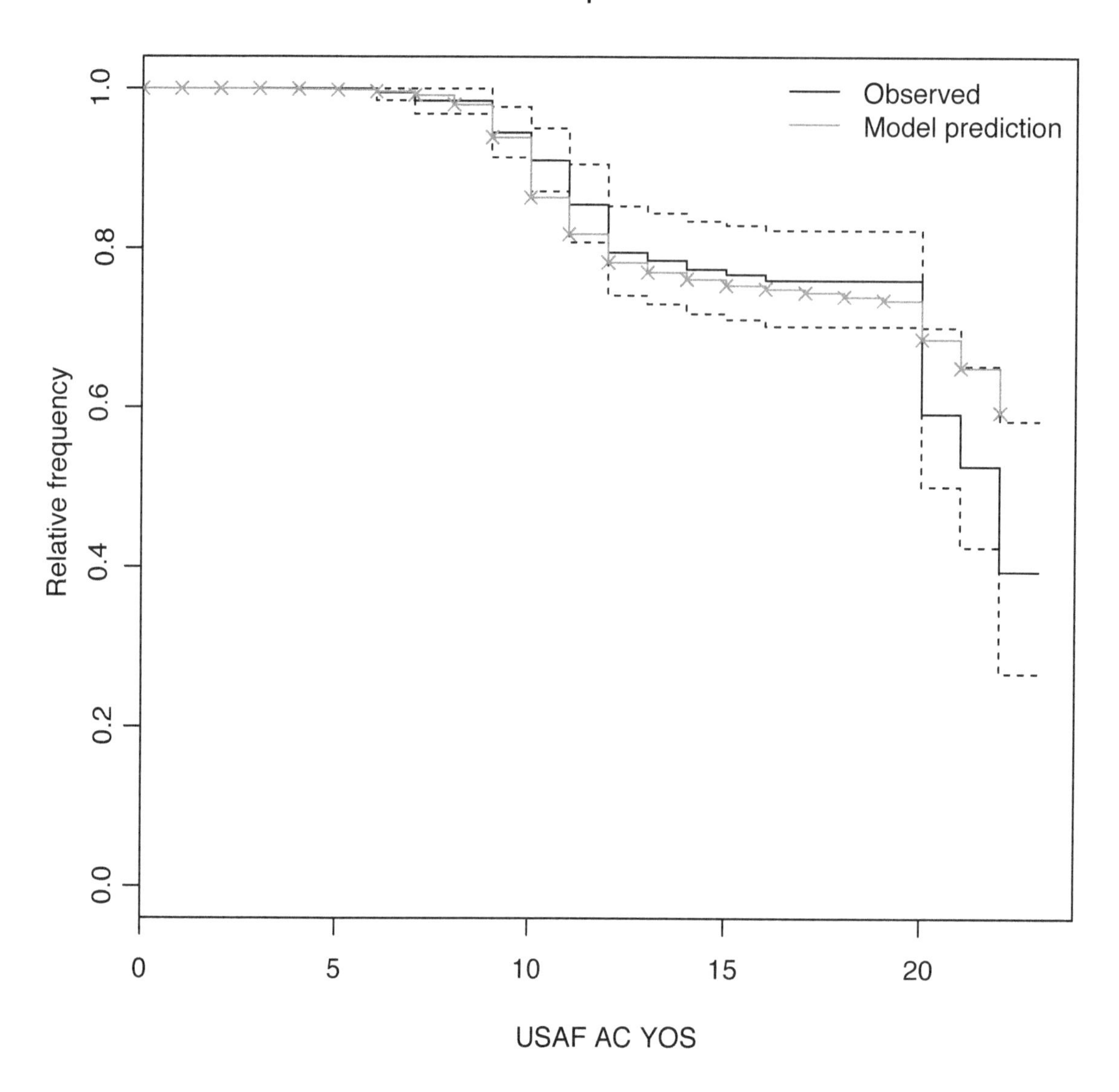

SOURCE: DMDC personnel data.
NOTE: The figure shows Kaplan–Meier cumulative probabilities of retention to each YOS in the AC. The black lines are the retention probabilities in the observed data, and the red lines are the predicted probabilities from the estimated models. The dotted lines are the error bands. For the most part, the red line is quite close to the black line and within the error band.

5. Results

The objective of our research was to estimate the efficient level of AvB by pilot community to meet and maintain a needed inventory level given the cost of accessing and retaining more pilots. We used the training cost estimates described in Chapter 3 together with the DRM estimates and the DRM computer coding discussed in Chapter 4 to develop a simulation capability that allowed us to simulate per capita pilot cost to the USAF as AvB increases. This chapter presents our simulation results.

Estimating Per Capita Cost

The formula for per capita cost is given by C in Chapter 2, or

$$C = \frac{PersonnelCost + \left(TrainingCostperPilot \times PilotsTrained\right)}{PilotForceSize}.$$

Training cost per pilot is given in the final column of Table 3.5 in Chapter 3. For example, the training cost per fighter pilot ranges from $5.6 million for an F-15E and F-16 pilot to $10.9 million for an F-22 pilot, in FY 2018 dollars. We computed the denominator in the formula, annual pilot force size for each community, using USAF Military Personnel Data System data that the Air Force Personnel Center provided. We used these data to compute the officer pilot inventory for each FY between 2008 and 2017. The pilot inventory has been changing over time, depending on the community; in our simulations, we used the median number of pilots, as shown in Table 5.1, for our estimate of pilot force size for each specialty and platform.

Table 5.1. Median Officer Pilot Inventory, Fiscal Years 2008 Through 2017

Pilot Type	Median Number of Pilots
Fighter	3,050
Mobility	815
Bomber	5,638
C2ISR	1,020

SOURCE: Military Personnel Data System data from the Air Force Personnel Center.

We used the DRM simulation capability to compute the number of pilots trained (i.e., the number of pilot accessions). We used the DRM estimates for each specialty and platform to simulate the cumulative retention probability by YOS or the percentage of entering pilots who reach each YOS. We then scaled the cumulative retention probabilities using the pilot force size figures in Table 5.1 so we could show the retention profile for each specialty and platform or the number of pilots by YOS. Given the simulated retention profile and the force size, we could then compute the number of or of pilots who need to be trained to sustain the pilot force size or inventory.

We also used the DRM simulation capability to simulate the personnel cost associated with a given retention profile. Personnel cost includes three elements:

- **AvB and AvIP costs**
- **regular military compensation:** This includes basic pay, basic allowance for housing, basic allowance for subsistence, and the tax advantage associated with receiving allowances tax-free
- **retirement accrual cost:** Military retirement is funded on an accrual basis, and DoD must contribute a percentage of its basic-pay bill to the Treasury. The percentage is called the normal cost percentage (NCP), and it is set by the DoD actuary each year. For the purpose of our simulations, we assumed an NCP of 29 percent, the NCP for FY 2017 for full-time personnel (DoD, Board of Actuaries, 2016).[6]

As AvB levels increase, the force becomes more experienced and the number of accessions required to sustain the force size falls, as shown in Figure 2.1 in Chapter 2. Consequently, the cost associated with AvB, regular military compensation, and the retirement accrual charge will increase. These changes will be reflected in the personnel cost in the computation of per capita pilot cost. Personnel cost is in FY 2018 dollars.[7]

We made several key assumptions in our simulations. First, we assumed that there were no training capacity or absorption constraints. Such constraints could increase the cost of training pilots because we would need to include the cost of additional capability. Thus, our simulations understated training cost insofar as these constraints are relevant. Second, we assumed that major-airline hiring was constant at 3,200 pilots per year in the steady state. We based this assumption on several factors. The number of pilots projected to retire from major airlines increases from about 1,000 in 2014 to more than 2,500 per year in 2022 through 2026, and nonretirement pilot attrition from major airlines is about 0.5 percent per year, or approximately 250 pilots per year (Mattock, Hosek, et al., 2016). In addition, the airlines are once again profitable and expanding their hiring, having emerged from a long period of bankruptcies and mergers. Reported major-airline hiring rose from 553 pilots in 2013 to 1,084 in 2014, 3,053 in 2015, 3,429 in 2016, and 4,113 in 2017 (FAPA, undated).

Simulation of How Steady-State Per Capita Cost Varies with the Aviation Bonus

We next show simulations of how steady-state cost per pilot varies with AvB.

---

[6] The NCP is for the legacy military retirement systems. Beginning in 2018, new military entrants are under a new military retirement system known as the Blended Retirement System; currently serving members are grandfathered into the legacy system, but personnel with 12 or fewer YOSs may opt into the new system.

[7] Specifically, we used the October 2017 Consumer Price Index for All Urban Consumers provided by the U.S. Bureau of Labor Statistics. Using the October 2017 index puts the personnel costs into the same FY 2018 dollars as the training costs.

Fighter Pilots

We begin with fighter pilots. As shown in the final column of Table 3.5 in Chapter 3, training cost per fighter pilot varies from \$5.6 million to \$10.9 million. Figure 5.1 shows per capita fighter pilot cost in thousands of dollars as the AvB cap for fighter pilots increases, assuming that training cost per pilot is either \$5.6 million or \$10.9 million.

Figure 5.1. Cost per Fighter Pilot, by Aviation Bonus Cap, When Training Cost per Pilot Ranges from \$5.6 Million to \$10.9 Million, in Thousands of Fiscal Year 2018 Dollars

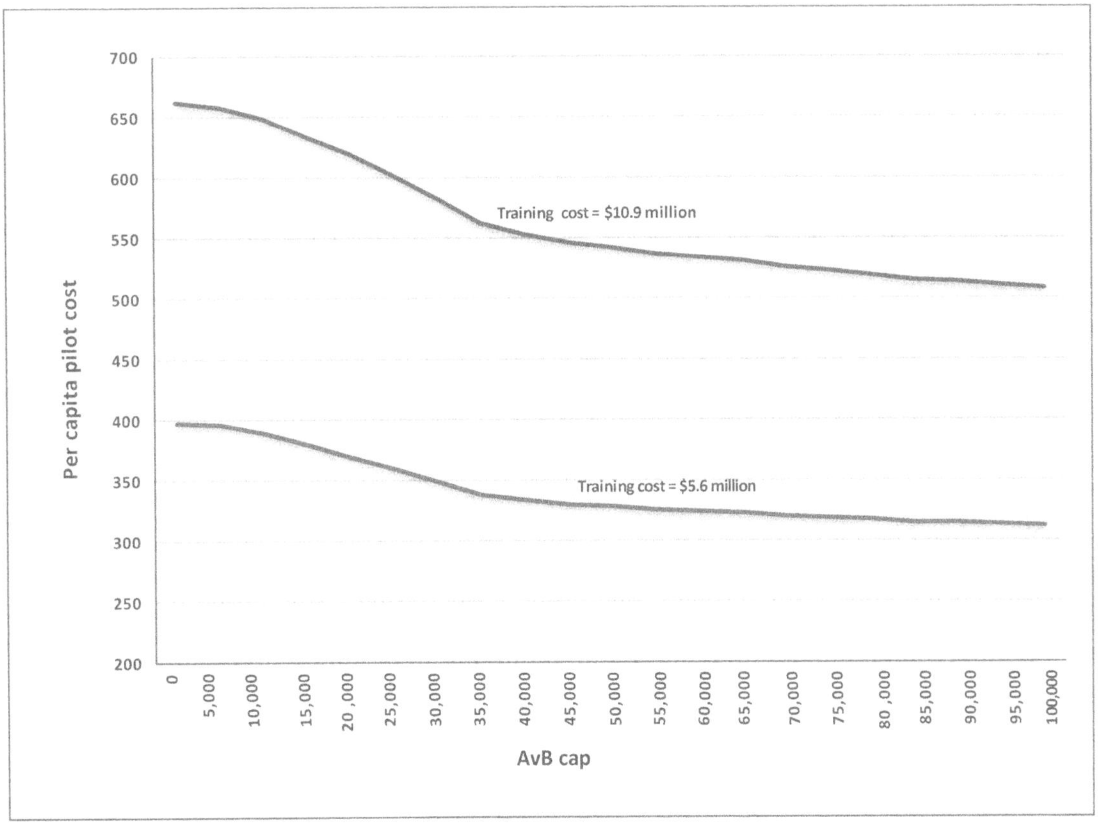

SOURCES: DRM simulation results; DMDC personnel data; AETC cost data.

The current AvB cap is \$35,000, but we were able to simulate the steady-state effects that an AvB cap that varies from \$0 to \$95,000 per year of commitment would have on retention, accessions, and per capita costs. Per capita cost is uniformly higher when training cost per pilot is higher. For example, at the current AvB cap of \$35,000, per capita cost is \$563,000 when training cost is \$10.9 million per pilot but \$339,000 when training cost per pilot is \$5.6 million. Note that per capita cost is in thousands of dollars (e.g., \$563,000) and not millions of dollars (\$10.9) because we computed per capita cost by dividing cost by the entire fighter pilot inventory.

The key finding in Figure 5.1 is that per capita fighter pilot cost is lower when AvB is higher for the range of per-pilot training costs we estimated for fighter pilots and for the range of AvB that we considered. In both cases—when training cost is $5.6 million per pilot and when it is $10.9 million per pilot—per capita cost declines. Put differently, per capita cost is lower when the USAF retains more pilots and accesses fewer pilots, even though personnel cost increases. The savings in training cost from making fewer accessions offsets the higher personnel cost associated with an inventory with more seniority. The decline in per capita cost for fighter pilots is nonlinear, with a more rapid decline when AvB is between $0 and $40,000 and a slower decline thereafter. The decline is nonlinear because the responsiveness of retention to increases in AvB is nonlinear. When AvB is low, pilot retention is more responsive to increases in AvB, but, as AvB increases and pilots with higher taste for the USAF are already retained, the remaining pilots are less responsive to a given increase of AvB. Consequently, at higher AvB levels, pilot retention is less responsive, yet cost increases. The result is that the decline in per capita cost is not as great at the highest levels of AvB we considered.

Because per capita fighter pilot cost is still decreasing at an AvB cap of $100,000 (the highest value we considered), the results indicate that, from the narrow standpoint of minimizing per capita pilot cost, it is efficient to increase the AvB cap to at least $100,000 and retain midcareer fighter pilots beyond the initial obligation. Figure 5.2 shows the steady-state fighter pilot retention profile when AvB is $25,000 (at baseline) versus $100,000. The experience mix of the fighter pilot force would be dramatically more senior if AvB were $100,000 than if AvB were lower and would be even more so if the AvB cap were even higher. Even at a cap of $100,000, nearly all pilots who complete their service obligations at YOS 11 would stay until at least YOS 20. And at an even higher AvB cap than $100,000, per capita cost would be lower, and virtually all fighter pilots would choose to stay after completing the initial obligation. Thus, the higher AvB leads to a more experienced force, a force that might be more experienced than the USAF wants. Specifically, the percentage of pilots with more than 20 YOSs increases when AvB is $100,000. Thus, whether it is efficient to continue to increase AvB also depends on USAF requirements for the most-experienced pilots.

Figure 5.2. Steady-State Fighter Pilot Cumulative Retention Profile, Baseline Versus an Aviation Bonus Cap of $100,000

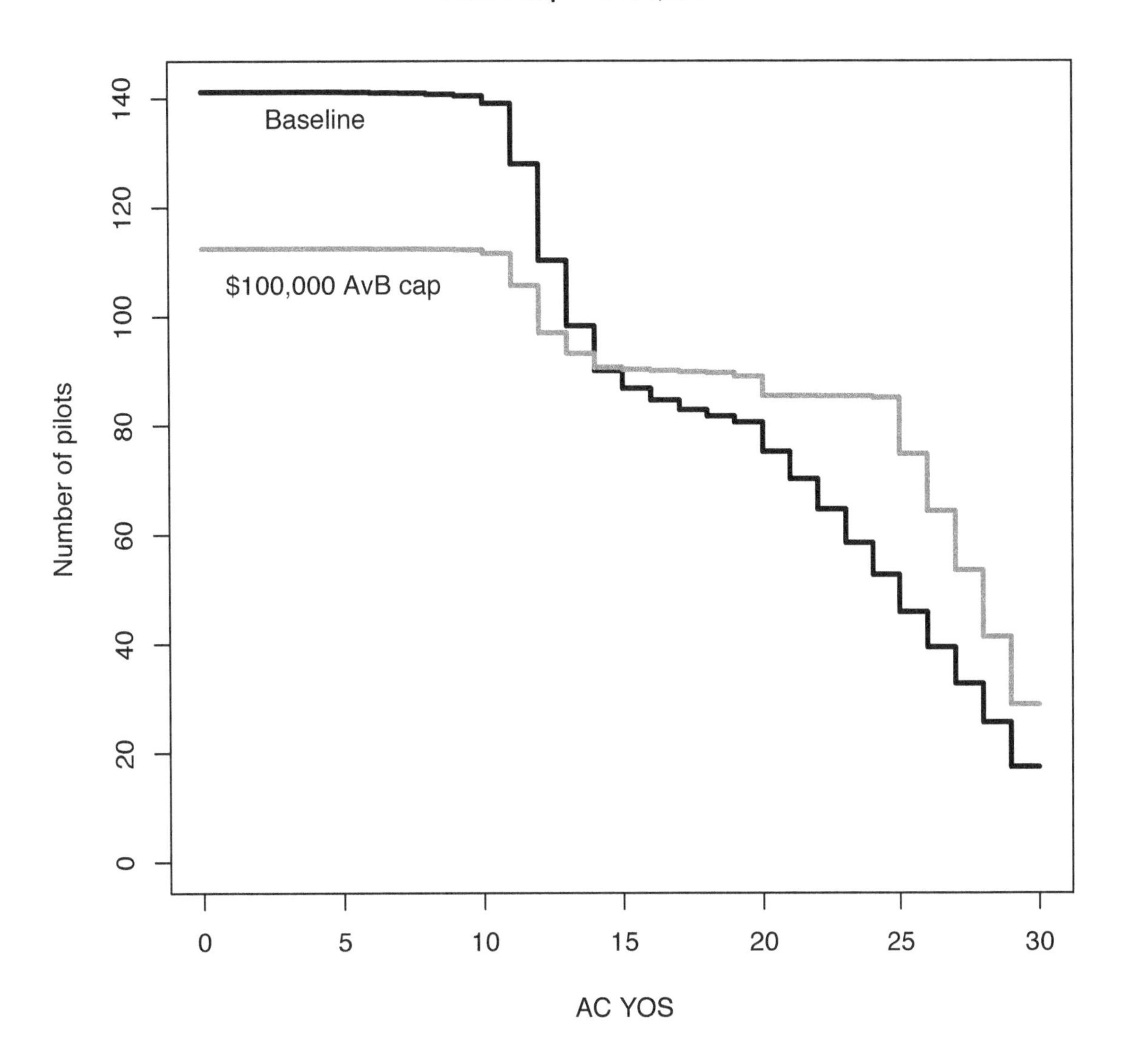

SOURCES: DRM simulation results; DMDC personnel data; AETC cost data.
NOTE: With increased AvB, there is no change in force, and the percentage of all pilots with 20 or more YOSs = –13.4; under the new policy, that percentage is 48.9.

The reason for our key finding that per capita cost for fighter pilots is lower when AvB is higher is that the cost of training a fighter pilot is so high, at least $5.6 million per pilot. Figure 5.3 illustrates this point. If the cost of training a fighter pilot were sufficiently low, per capita cost would not continue to decline as AvB increases. For example, if training cost per fighter pilot were $1 million per pilot or less, per capita cost would be minimized when the AvB cap was $45,000. But, at the USAF's current high training cost for fighter pilots, continuing to increase AvB is efficient in terms of minimizing per capita cost to continue to retain fighter pilots—subject to the caveat about the USAF's preferred experience mix.

Figure 5.3. Cost per Fighter Pilot, by Aviation Bonus Cap, When Training Cost per Pilot Ranges from $550,000 to $3 Million, in Thousands of Fiscal Year 2018 Dollars

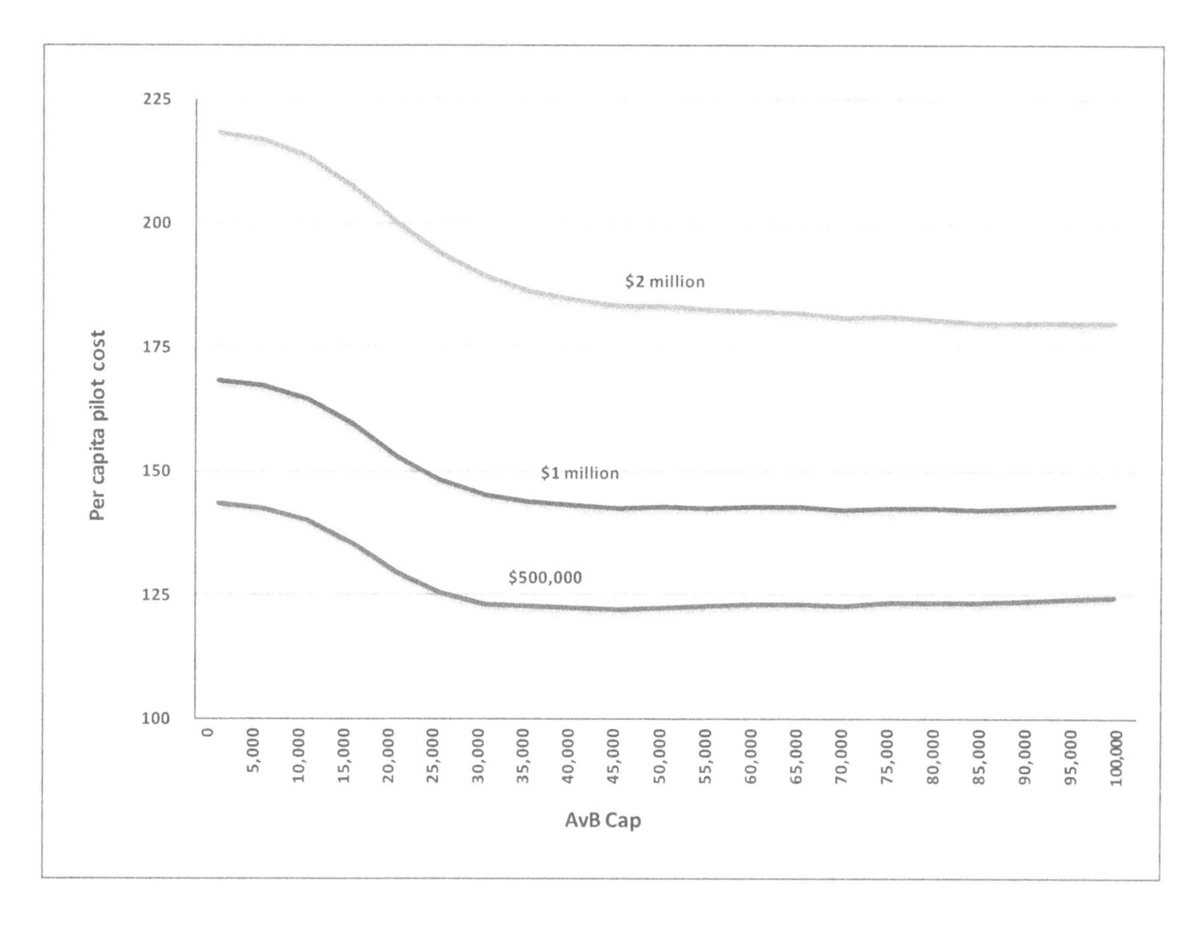

SOURCES: DRM simulation results; DMDC personnel data; AETC cost data.

Mobility; Command, Control, Intelligence, Surveillance, and Reconnaissance; and Bomber Pilots

We found qualitatively similar results for mobility and C2ISR pilots as for fighter pilots, as shown in Figures 5.4 and 5.5, respectively. For the range of per-pilot training costs that we describe in Chapter 3, we found that per capita pilot cost decreases as AvB increases.

Figure 5.4 shows results for when per capita training cost for mobility pilots ranges from $1.1 million, the approximate costs for C-17 and KC-135 pilots, to $2.5 million, the per-pilot training cost for C-130J pilots. For C-130J pilots, as AvB increases, per capita mobility pilot cost decreases over the range we considered. As AvB increases, retention increases and required accessions to sustain the pilot force decreases. Training cost decreases because fewer pilots are trained, but personnel cost increases as the pilot force becomes more experienced. The decrease in training cost offsets the increase in personnel cost, so overall per capita cost decreases as AvB increases. As before, the decrease is nonlinear, being larger when AvB is smaller and smaller when AvB is larger. As AvB increases, it becomes less effective at the margin and induces fewer pilots to stay. Also, like before, the experience mix increases as AvB increases, perhaps beyond the point that the USAF finds desirable. In the case of C-17 pilots (and KC-135 pilots, not shown), per capita cost decreases as AvB increases from $0 to about $40,000, whereupon per capita cost declines less rapidly. In this case, the higher personnel cost is offset less and less by lower training cost, so, on a per capita basis, cost declines less but does not rise.

Figure 5.4. Cost per Mobility Pilot, by Aviation Bonus Cap, When Training Cost per Pilot Ranges from $1.1 Million to $2.5 Million, in Thousands of Fiscal Year 2018 Dollars

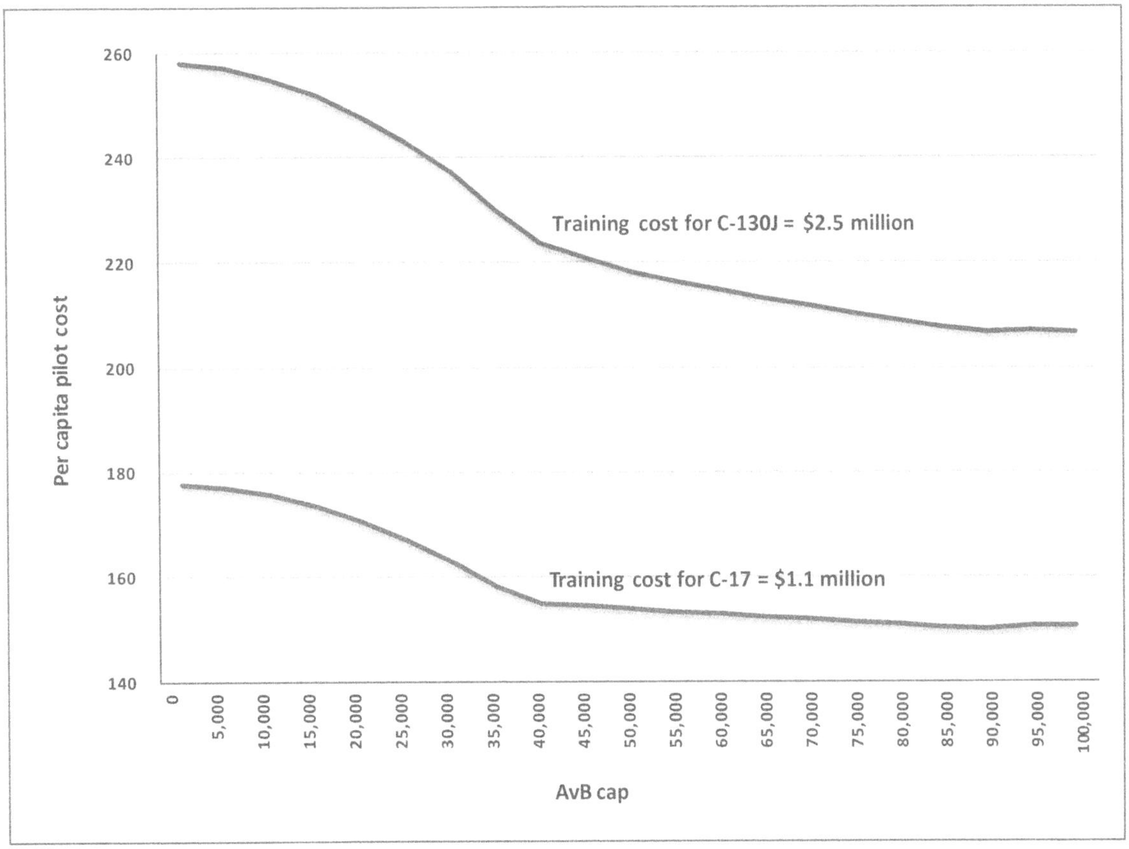

SOURCES: DRM simulation results; DMDC personnel data; AETC cost data.

We estimated per-pilot training cost to be $5.4 million for RC-135 pilots. We found that C2ISR per capita cost decreases steadily as AvB increases (Figure 5.5) over the range of AvB cap that we considered. As discussed in the context of Figure 4.1 in Chapter 4, the C2ISR community is more responsive to changes in major-airline hiring because its pilots tend to accumulate more flying hours. Even when the AvB cap is as high as $100,000, per capita pilot cost still declines, unlike costs for the other communities we considered, such as mobility pilots in Figure 5.4, for whom per capita cost levels out because an increase in the AvB cap has a larger effect on retention and accessions for the C2ISR community at high levels of AvB than for other communities.

Figure 5.5. Cost per Command, Control, Intelligence, Surveillance, and Reconnaissance Pilot, by Aviation Bonus Cap, When Training Cost per Pilot Is $5.4 Million, in Thousands of Fiscal Year 2018 Dollars

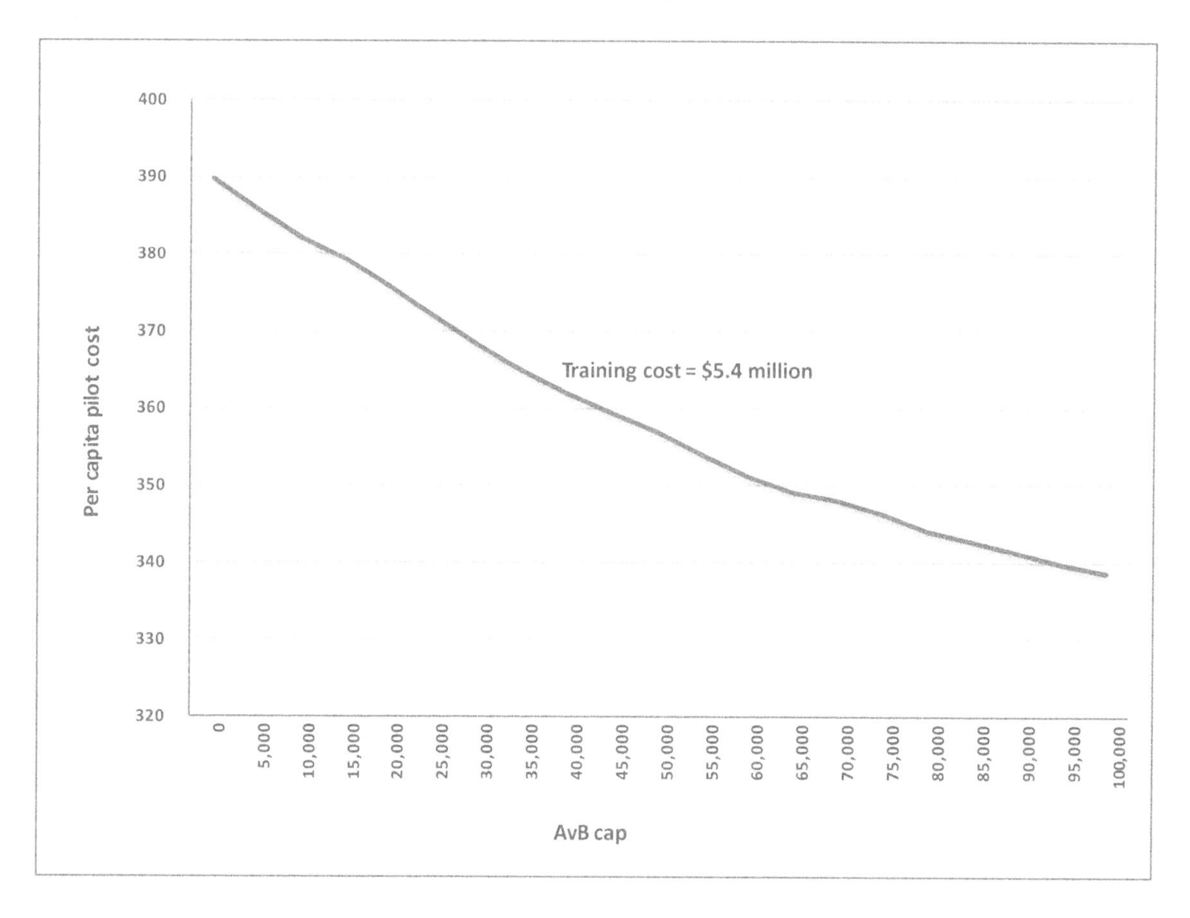

SOURCES: DRM simulation results; DMDC personnel data; AETC cost data.

For bomber pilots, per-pilot training costs range from $7.3 million for B-1 pilots to $9.9 million for B-2 pilots, but the results for bomber pilots differ somewhat from those for the other pilot communities, as shown in Figure 5.6. Like with the other communities, per capita cost declines as the AvB cap increases from $15,000 to $45,000, but per capita cost then increases until the cap is $55,000 and then decreases again until the cap is $75,000. Beyond a cap of $75,000, per capita cost is relatively stable as AvB increases. This unusual pattern for bomber pilots is due to their retention rate being much higher than of other pilot communities, as shown in Figure 4.6 in Chapter 4. As a result, increases in the AvB cap from $45,000 to $55,000 or $60,000 have relatively little retention effect on midcareer pilots because their retention rate is already very high; the main effect is to induce more senior pilots to defer retirement. It is not until the AvB is increased to $65,000 or more that the predominant effect returns to inducing midcareer pilots to stay.

Figure 5.6. Cost per Bomber Pilot, by Aviation Bonus Cap, When Training Cost per Pilot Ranges from $7.3 Million to $9.9 Million, in Thousands of Fiscal Year 2018 Dollars

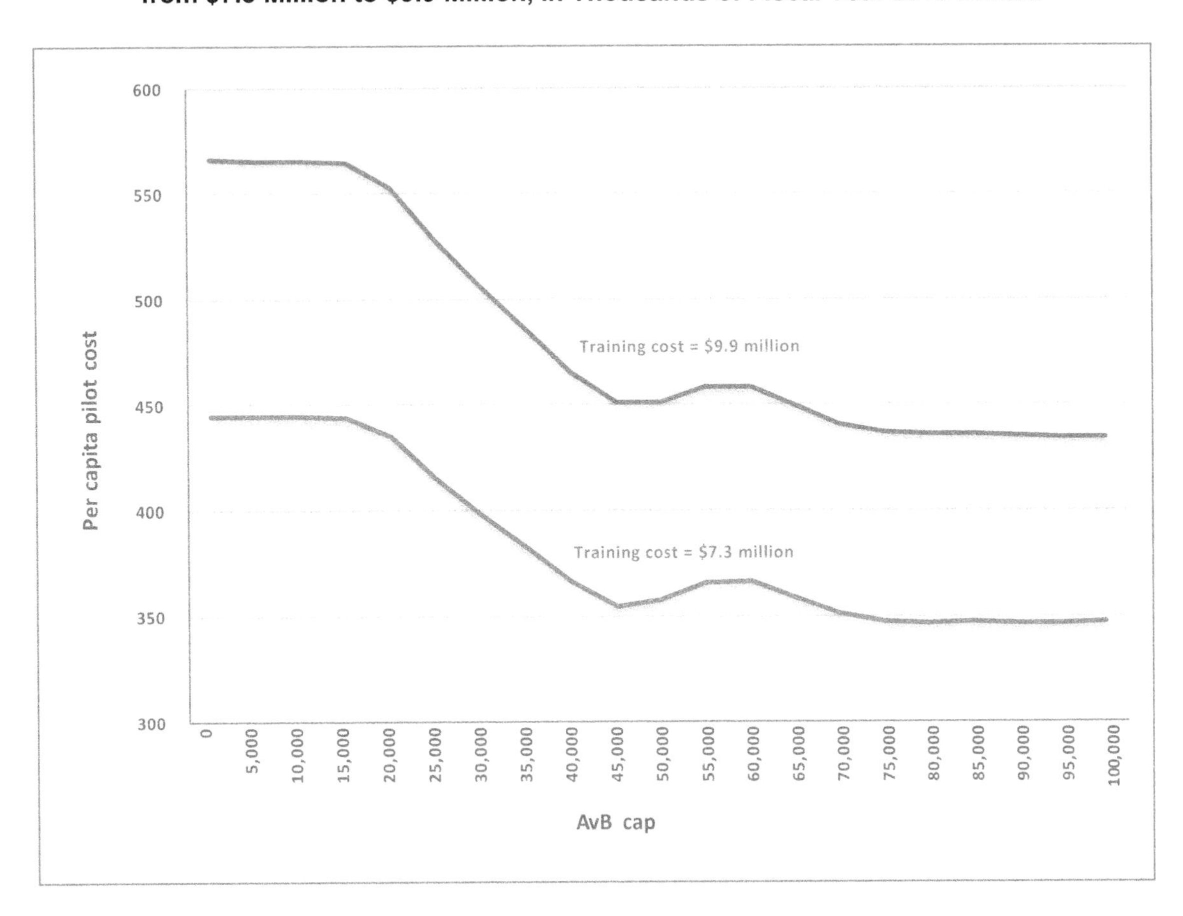

SOURCES: DRM simulation results; DMDC personnel data; AETC cost data.

Simulations of the Transition to the Steady State

The results shown so far are for the steady state. That is, we showed how per capita cost varies when AvB versus accessions vary, when enough time has passed so that all members have spent their entire careers under a new regime of higher AvB or higher accessions. It is also of interest to consider short-term effects during the transition to the new steady state. Because of the amount of time required to train an experienced pilot, changing the force size—say, by increasing accessions rather than increasing retention—can increase the pilot inventory in the steady state but could mean delays in the short run because time is required before the new pilot accessions complete the training pipeline. Consequently, in the transition years, the number of trained pilots could differ substantially, depending on whether the USAF increases the force size by increasing accessions or does so by increasing AvB and therefore retention.

In this section, we provide estimates that illustrate this point by considering an example in which the USAF raises AvB versus raises accessions to increase the size of the pilot force. We show the steady-state and transitional effects on the experience mix of the fighter pilot force in each case. We first show that, in the steady state, increasing the AvB cap from $25,000 to

$35,000 or, alternatively, increasing accessions can achieve an overall increase in force size of 6.5 percent. These cases are extreme because, in reality, the USAF would likely increase both accessions and retention to increase the force size, but they illustrate the point that, in the transition years, increasing the inventory of experienced pilots takes longer when the force is expanded by accessions rather than by increasing retention.

Figure 5.7 shows the steady-state effects of increasing the fighter pilot force by 6.5 percent by increasing AvB from $25,000 to $35,000 (right) versus increasing accessions by 6.5 percent (left). The black line is the steady-state baseline fighter force profile by YOS when AvB is $25,000; the red line is the steady-state fighter force profile after the increase in AvB (right) or a change in accessions (left). For the base case, we assume the fighter pilot force to be 3,050. In the new steady state, the pilot force increases by 6.5 percent to 3,248. For both the base case and the alternatives, we assumed that major-airline hiring was 3,100 per year.

Figure 5.7. Steady-State Fighter Pilot Retention Profile with an Increase in Accessions or in Aviation Bonus

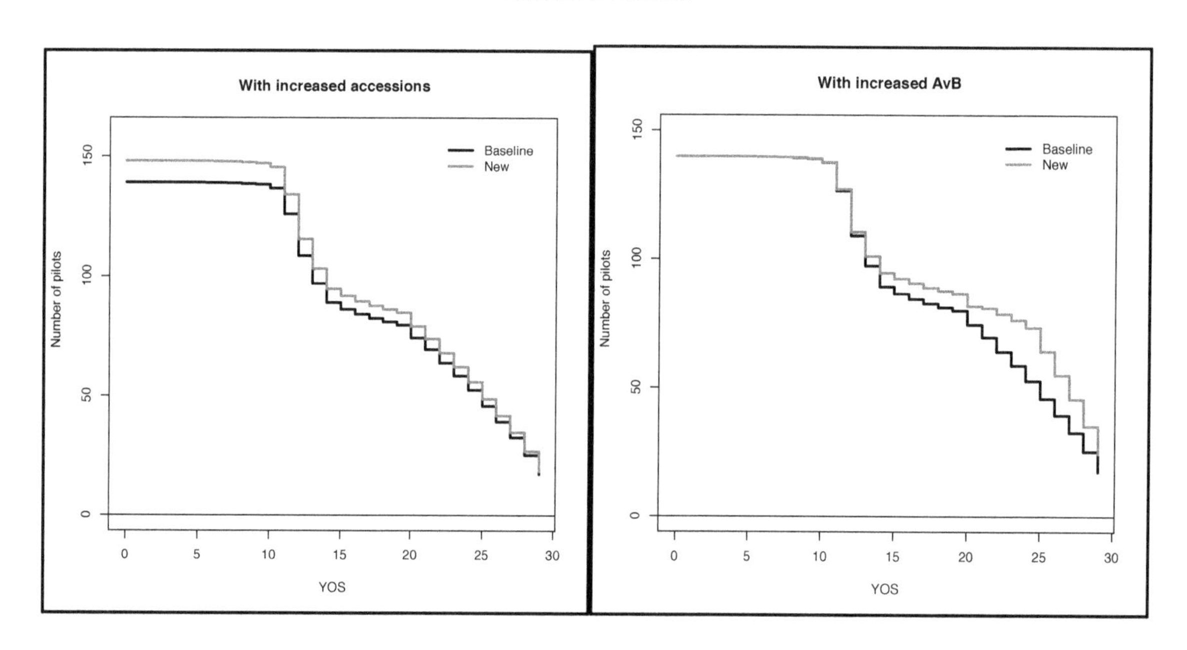

SOURCES: DRM simulation results; DMDC personnel data; AETC cost data.
NOTE: With increased accessions (left panel), the change in force is 6.5 and the percentage of all pilots with 12 or more YOSs = 46.2; under the new policy, that percentage is 46.2. With increased AvB (right panel), the percentages are 46.2 and 49.4, respectively.

The panels in Figure 5.7 show that, when the fighter pilot force size is increased by increasing AvB, the force becomes more senior, with more pilots staying in the USAF beyond 11 YOSs in the steady state, while the experience mix in the steady state stays the same when the force is increased by increasing accessions. The note beneath each panel shows the percentage of all fighter pilots who are midcareer and experienced, defined as the percentage of pilots with 12 or more YOSs. We chose 12 YOSs because fighter pilots complete their initial service obligations after 11 YOSs, and this is the first key retention decision point for trained and experienced pilots. In the baseline, the percentage is 46.2. In the new steady state, the percentage is also 46.2 when accessions are increased; the percentage is 49.4 when AvB is increased instead.

In Figures 5.8 through 5.10, we show how the share of experienced pilots varies in the transition to the steady state when the number of accessions rather than AvB is increased. We show results for $t = 5$, 12, and 17, where t is the number of years after the change in AvB or accessions. We assumed that the increase in AvB or in accessions is permanent. In each figure, the left panel shows the force profile by YOS when accessions are increased by 6.5 percent and the right panel shows the profile when the AvB cap is increased instead to $35,000 to achieve a 6.5-percent force size increase and accessions are held constant.

Figure 5.8 shows the force profiles at $t = 5$ or five years after accessions or, alternatively, AvB is increased. When accessions are increased (Figure 5.8, left panel), the number of pilots with five or fewer YOSs increases (left panel), meaning that more pilots are in the training pipeline. But without any increase in the number of more-senior pilots, the share of experienced pilots is lower (45.5 percent) than in the steady state. After 12 years (Figure 5.9, left panel), those who entered the pipeline in year 1 are now experienced midcareer pilots who have completed the 11-year service obligation and are eligible to leave. Increasing accessions has increased the pool of pilots making the first key retention decision, but it takes 12 years to do so. Figure 5.10 (left panel) shows the expanded pool of experienced midcareer pilots after 17 years. Eventually, the pilots in the larger pipeline gain experience so that, in the new steady state, the pilot force grows proportionately at each YOS and the share of pilots who are experienced is 46.2 percent, like in the baseline.

In contrast, when the USAF raises the AvB cap to $35,000 while keeping the number of accessions constant, the share of pilots who are trained and experienced and have at least 12 YOSs is higher than when accessions are increased. After five years (Figure 5.8, right panel), the share remains constant at 46.2 percent, in contrast to the drop in the share after five years when accessions are increased instead. After 12 years, the share increases to 48.4 percent (Figure 5.9, right panel), greater than the share when accessions are increased and greater than in the base case. Increasing Aviator Retention Pay (ARP) without increasing accessions does not increase the training pipeline, as is the case when accessions are increased, but experienced midcareer pilots stay longer, so the share of these pilots is higher. After 17 years, the share increases even more, to 49 percent (Figure 5.10, right panel), in contrast to the case when accessions are increased.

Although we do not show the results, we found similar qualitative results for the other pilot communities.

Figure 5.8. Transition Fighter Pilot Retention Profile at t = 5 with an Increase in Accessions or in Aviation Bonus

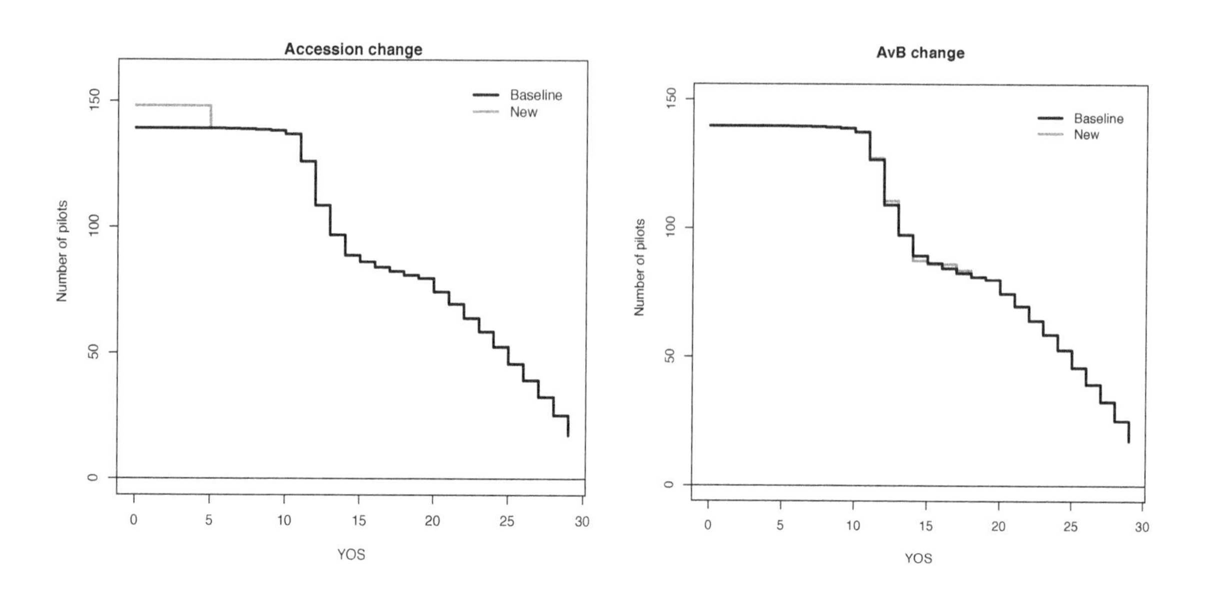

SOURCES: DRM simulation results; DMDC personnel data.

NOTE: When accessions are increased (left panel), the number of pilots with five or fewer YOSs increases, meaning that more pilots are in the training pipeline. But without any increase in the number of more-senior pilots, the share of pilots who are experienced is lower (45.5 percent) than in the steady state. In contrast, when the USAF raises the AvB cap to $35,000 while keeping the number of accessions constant, the share of pilots who are trained and experienced and have at least 12 YOSs is higher than when accessions are increased. After five years (right panel), the share remains constant at 46.2 percent, in contrast to the drop in the share after five years when accessions are increased instead. Although we do not show the results, we found similar qualitative results for the other pilot communities.

Figure 5.9. Transition Fighter Pilot Retention Profile at *t* = 12 with an Increase in Accessions or in Aviation Bonus

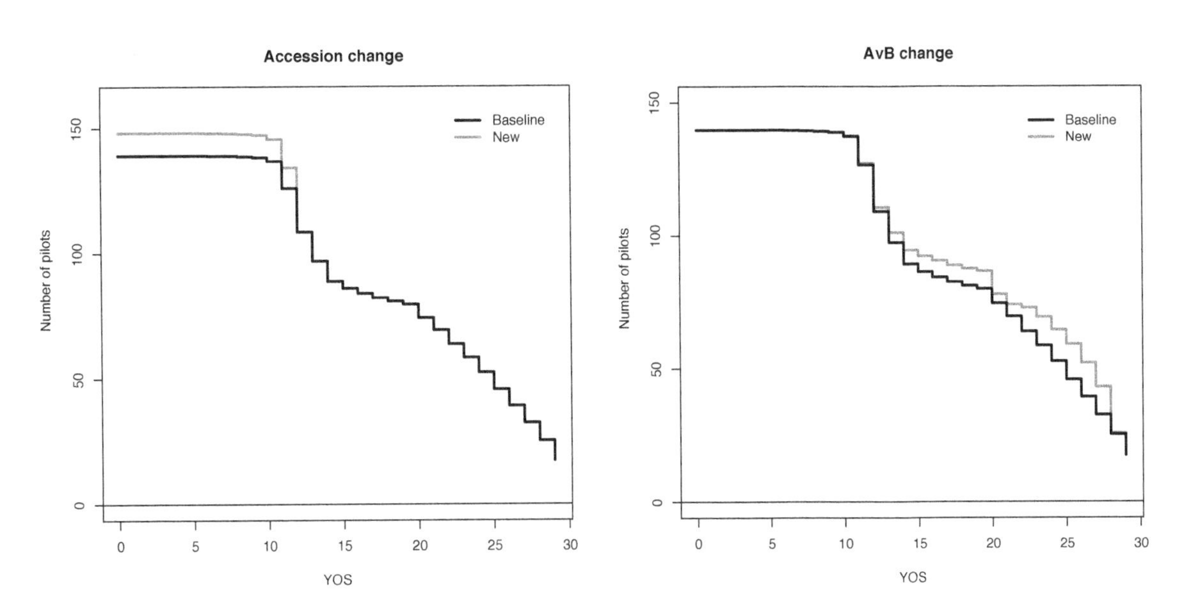

SOURCES: DRM simulation results; DMDC personnel data.

NOTE: After 12 years (left), those who entered the pipeline in year 1 are now experienced midcareer pilots who have completed the 11-year service obligation and are eligible to leave. Increasing accessions has increased the pool of pilots making the first key retention decision, but it takes 12 years to do so. In contrast, when the USAF raises the AvB cap to $35,000 while keeping the number of accessions constant, the share of pilots who are trained and experienced and have at least 12 YOSs is higher than when accessions are increased. After 12 years, the share increases to 48.4 percent (right), greater than the share when accessions are increased and greater than in the base case. Increasing ARP without increasing accessions does not increase the training pipeline, as is the case when accessions are increased, but experienced midcareer pilots stay longer, so the share of these pilots is higher. Although we do not show the results, we found similar qualitative results for the other pilot communities.

Figure 5.10. Transition Fighter Pilot Retention Profile at $t = 17$ with an Increase in Accessions or in Aviation Bonus

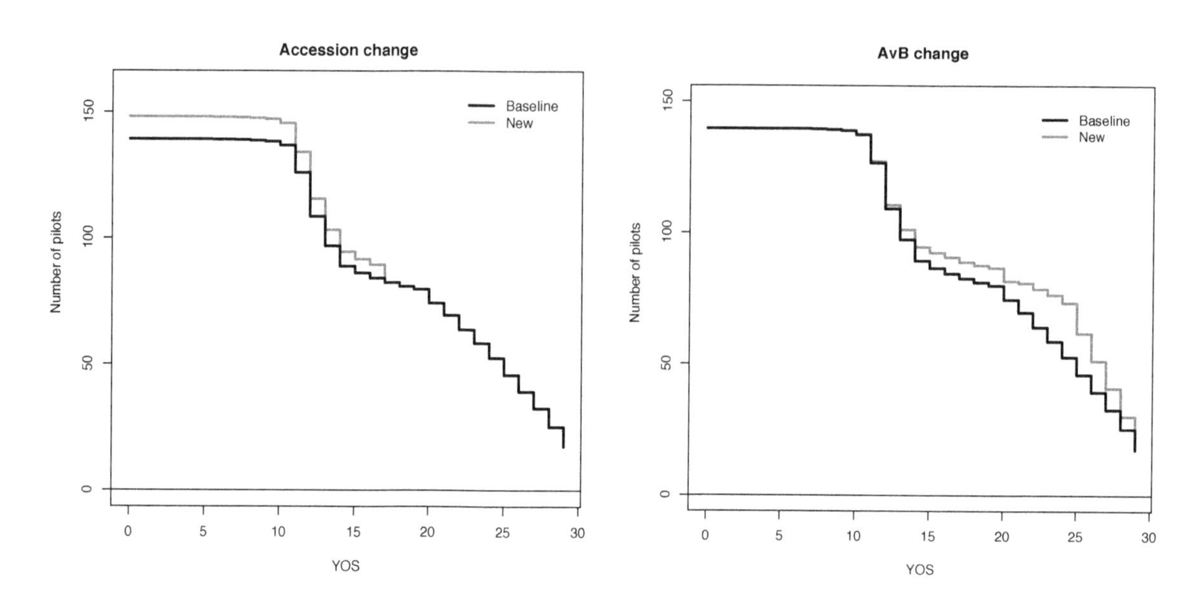

SOURCES: DRM simulation results; DMDC personnel data.

NOTE: The left panel shows the expanded pool of experienced midcareer pilots after 17 years. Eventually, the pilots in the larger pipeline gain experience, so in the new steady state, the pilot force grows proportionately at each YOS and the share of experienced pilots is 46.2 percent, like in the baseline. In contrast, when the USAF raises the AvB cap to $35,000 while keeping the number of accessions constant, the share of pilots who are trained and experienced and have at least 12 YOSs is higher than when accessions are increased. After 17 years, the share increases even more, to 49 percent (right), in contrast to the case when accessions are increased. Although we do not show the results, we found similar qualitative results for the other pilot communities.

6. Concluding Thoughts

The key conclusion of our research is that increasing AvB to increase retention is more efficient than expanding the pipeline to sustain a given pilot inventory. This result holds for all of the communities we considered: fighter, bomber, C2ISR, and mobility pilots. Our analysis did not incorporate absorption or training capacity constraints that might be relevant when the training pipeline is expanded, but accounting for these constraints would make increasing S&I pay even more attractive in terms of avoiding the cost of increasing capacity and higher training cost. The proviso we attach to this result is that, depending on its size, the increase in AvB might result in a steady-state pilot inventory that is more senior than the USAF wants.

Another key conclusion of our research is that it takes longer to increase the inventory of experienced pilots, especially in the midcareer, when pilots are free to make retention decisions, when the force is expanded by accessions rather than by increasing S&I pay and therefore retention. Because training pilots takes time, increasing the training pipeline as a means of increasing the pilot inventory would also mean that the share of pilots who are experienced will be lower in the initial transition years than if the USAF were to increase S&I pay instead.

Increasing AvB is more efficient than increasing accessions because the former policy results in a more experienced force that requires fewer accessions to sustain a given force size. This more experienced force is also more expensive in terms of personnel cost, but the savings from training fewer pilots offset this cost. The savings in training cost are significant because of the high cost of training a given pilot. We estimated that the training cost per qualified fighter pilot is between \$5.6 million and \$10.9 million, depending on the platform. The cost of training a bomber pilot is between \$7.3 million and 9.9 million per pilot. Flying hours and the high cost per flying hour of these platforms explain a large share of this training cost. At training cost this high, our result that increasing AvB is more efficient is not sensitive to the precision of the estimate of training cost. For example, even if training cost were half of what we estimated for fighter pilots, we would still find that increasing AvB is at least as efficient, if not more efficient, than increasing accessions to sustain a given pilot inventory.

In short, our results about the efficiency of AvB over accessions are driven by the high cost of training pilots. This training cost is determined by the current curricula, policies, capacities, and technologies used to train USAF pilots. Changes in how pilots are trained could potentially impart savings, although the changes would have to be sufficiently large to reverse our results and make expanding the training pipeline more cost-effective than expanding AvB. Given that flying hours and the cost per flying hour are important drivers of training cost, reducing flying hours—say, through simulator technology—could be one area in which sufficient cost savings could occur. However, to maintain capability, it would be important that simulators provide the same quality of training and experience as actual flying. Future research should explore where potential training cost savings might be found and whether alternative technologies might provide a source of savings without jeopardizing USAF capability and readiness.

Bibliography

AETC—*See* Air Education and Training Command.

Air Combat Command, "USAF Operations Training Course: B-1 Basic Qualification Course," course syllabus, Joint Base Langley–Eustis, Va., September 2014a, not available to the general public.

———, "USAF Operations Training: F-22 Basic Qualification, Transition/Requalification, Instructor Pilot Training Courses," course syllabus, Joint Base Langley–Eustis, Va., January 2014, incorporating change 2, November 2014b, not available to the general public.

———, "USAF Operations Training: F-15E Basic Qualification Training Course," course syllabus, Joint Base Langley–Eustis, Va., January 2015, not available to the general public.

———, "USAF Operations Training: RC-135 Pilot Qualification and Requalification," course syllabus, Joint Base Langley–Eustis, Va., June 2016, not available to the general public.

———, "USAF Operations Training: A-10C Pilot Initial Qualification Training Course," course syllabus, Joint Base Langley–Eustis, Va., April 2016, incorporating change 1, January 2017, not available to the general public.

Air Education and Training Command, "Flying Training," undated. As of October 25, 2017: http://www.aetc.af.mil/flying-training/

———, "Flying Training: T-38C Specialized Undergraduate Pilot Training," course syllabus, Randolph Air Force Base, Texas, October 2012, not available to the general public.

———, "Flying Training: USAF Introduction to Fighter Fundamentals (IFF) T-38C," course syllabus, Randolph Air Force Base, Texas, April 2012, through change 2, April 2013, not available to the general public.

———, "Flying Training: C-17 Pilot Initial Qualification (PIQ) (FMS included) AC 15-1," course syllabus, Randolph Air Force Base, Texas, January 2014a, not available to the general public.

———, "Flying Training: C-130J Pilot Initial Qualification, C-130J Pilot Non-MAF Initial Qualification, C-130J Pilot Transition Long (Qualification), C-130J Pilot Transition Short (Qualification), C-130J Pilot Non-MAF Transition Short (Qualification), C-130J Pilot Requalification Qualification), C-130J Senior Officer Course Initial Qualification (Restricted)," course syllabus, Randolph Air Force Base, Texas, October 2014b, not available to the general public.

———, "Flying Training: C-130J Pilot Mission Qualification, C-130J Pilot Non-MAF Mission Qualification, C-130J Pilot Transition Long (Mission), C-130J Pilot Transition Short

(Mission), C-130J Pilot Non-MAF Transition Short (Mission), C-130J Pilot Requalification (Mission)," course syllabus, Randolph Air Force Base, Texas, October 2014c, not available to the general public.

———, *Performance Work Statement (PWS) for Initial Flight Training (IFT) Re-Competition*, December 4, 2014d.

———, "Flying Training: F-16C/D Initial Qualification (56 FW), F-16C/D Requalification (56 FW), F-16C/D Specialized Qualification (56 FW)," course syllabus, Randolph Air Force Base, Texas, September 2015a, not available to the general public.

———, "Flying Training: USAF F-15 Initial Qualification Course, USAF F-15 Transition/Requalification Course, USAF F-15 Senior Officer/Test Pilot Course," course syllabus, Randolph Air Force Base, Texas, October 2015b, not available to the general public.

———, "Flying Training: KC-135 Pilot Initial Qualification, KC-135 Pilot Transition Course 1, 2, and 3 (Qualification)," course syllabus, Randolph Air Force Base, Texas, November 2015c, not available to the general public.

Air Force Global Strike Command, "USAF Operations Training B-2A Initial Qualification Training," course syllabus, Barksdale Air Force Base, La., April 2012.

———, "USAF Operations Training B-52 Pilot Initial Qualification Course," course syllabus, Barksdale Air Force Base, La., March 2016, not available to the general public.

Air Force Total Ownership Cost, homepage, last modified July 26, 2018.

Air Mobility Command, "C-5 Aircrew Training System Pilot Initial Qualification Course," course syllabus, Scott Air Force Base, Ill., August 2016.

Anders, Earl, operations research analyst, Accounting and Cost Branch, Headquarters, Air Education and Training Command/Financial Management and Comptroller, "Process to Capture Pilot Training Costs," email to author Michael Boito, May 3, 2017a.

———, "Surge Pilot Production," email to author Michael Boito, October 31, 2017b.

Asch, Beth J., James Hosek, Michael G. Mattock, and Christina Panis, *Assessing Compensation Reform: Research in Support of the 10th Quadrennial Review of Military Compensation*, Santa Monica, Calif.: RAND Corporation, MG-764-OSD, 2008. As of September 20, 2018: https://www.rand.org/pubs/monographs/MG764.html

Department of the Air Force, *Fiscal Year (FY) 2018 Budget Estimates: Operations and Maintenance, Air Force*, Vol. I, May 2017. As of August 9, 2018: https://www.saffm.hq.af.mil/Portals/84/documents/Air%20Force%20Operation%20and%20Maintenance%20Vol%20I%20FY18.pdf?ver=2017-05-23-154654-623

DoD—*See* U.S. Department of Defense.

FAPA—*See* Future and Active Pilot Advisors.

Future and Active Pilot Advisors, "Major Airline Pilot Hiring by Year (2000–Present)," undated. As of August 13, 2018:
http://fapa.aero/hiringhistory.asp

Hosek, James, Shanthi Nataraj, Michael G. Mattock, and Beth J. Asch, *The Role of Special and Incentive Pays in Retaining Military Mental Health Care Providers*, Santa Monica, Calif.: RAND Corporation, RR-1425-OSD, 2017. As of September 20, 2018:
https://www.rand.org/pubs/research_reports/RR1425.html

Laverson, Alan, *A Study of Overhead Rate Behavior at a U.S. Air Force Base in the Context of A-76 Competitions*, Santa Monica, Calif.: RAND Corporation, RGSD-150, 2000. As of August 9, 2018:
https://www.rand.org/pubs/rgs_dissertations/RGSD150.html

Logistics, Installations and Mission Support—Enterprise View, website, undated, not available to the general public.

Mattock, Michael G., and Jeremy Arkes, *The Dynamic Retention Model for Air Force Officers: New Estimates and Policy Simulations of the Aviator Continuation Pay Program*, Santa Monica, Calif.: RAND Corporation, TR-470-AF, 2007. As of November 30, 2017:
http://www.rand.org/pubs/technical_reports/TR470.html

Mattock, Michael G., James Hosek, Beth J. Asch, and Rita Karam, *Retaining U.S. Air Force Pilots When the Civilian Demand for Pilots Is Growing*, Santa Monica, Calif.: RAND Corporation, RR-1455-AF, 2016. As of November 30, 2017:
https://www.rand.org/pubs/research_reports/RR1455.html

McDonald, William, then chief, Rated Force Policy, Military Force Policy Division, U.S. Air Force, "RFI for RAND Study," email to the authors, September 14, 2017.

McGee, Michael, *Air Transport Pilot Supply and Demand: Current State and Effects of Recent Legislation*, Santa Monica, Calif.: RAND Corporation, RGSD-351, 2015. As of November 30, 2017:
http://www.rand.org/pubs/rgs_dissertations/RGSD351.html

Office of the Under Secretary of Defense for Personnel and Readiness, "Special and Incentive Pay Index: Title 37, Chapter 5, Subchapter I—S&I Pays Currently for Active Duty Members," undated. As of August 10, 2018:
https://militarypay.defense.gov/Pay/Special-and-Incentive-Pays/Index/

———, *Aviation Incentive Pays and Bonus Program*, Department of Defense Instruction 7730.67, October 20, 2016. As of November 30, 3017:
http://www.esd.whs.mil/Portals/54/Documents/DD/issuances/dodi/773067_dodi_2016.pdf

Sweeney, Nolan, *Predicting Active Duty Air Force Attrition Given an Anticipated Increase in Major Airline Pilot Hiring*, Santa Monica, Calif.: RAND Corporation, RGSD-338, 2015. As of May 17, 2018: https://www.rand.org/pubs/rgs_dissertations/RGSD338.html

U.S. Bureau of Labor Statistics, "Consumer Price Index: October 2017," news release, USDL-17-1503, 2017. As of September 20, 2018: https://www.bls.gov/news.release/archives/cpi_11152017.pdf

U.S. Department of Defense, "Contracts for Jan. 13, 2017," press release CR-009-17, January 13, 2017. As of October 25, 2017: https://www.defense.gov/News/Contracts/Contract-View/Article/1049936/

U.S. Department of Defense, Board of Actuaries, *2016 Report to the President and Congress*, December 2016.

Woody, Christopher, "The Air Force May Pay Pilots Nearly a Half-Million Dollars to Stay in Uniform," *Business Insider*, March 30, 2017. As of August 10, 2018: https://www.businessinsider.com/air-force-pay-pilots-bonus-incentives-to-stay-in-uniform-2017-3

CPSIA information can be obtained
at www.ICGtesting.com
Printed in the USA
LVHW062047231019
635121LV00013B/132/P

9 781977 402042